Theory of Sets

BY DR. E. KAMKE

PROFESSOR OF MATHEMATICS, UNIVERSITY OF TÜBINGEN

TRANSLATED BY FREDERICK BAGEMIHL

ASSISTANT PROFESSOR OF MATHEMATICS, UNIVERSITY OF ROCHESTER

NEW YORK

Dover Publications, Inc.

This Dover edition, first published in 1950, is an English translation of the second German edition of *Mengenlehere* by E. Kamke.

Standard Book Number: 486-60141-2

Library of Congress Catalog Card Number: 50-7144

Manufactured in the United States by Courier Corporation
60141220
www.doverpublications.com

Contents

Introduction

THE THEORY OF SETS, which was founded by G. Cantor (1845–1918) and already developed by him into an admirable system, is one of the greatest creations of the human mind. In no other science is such bold formation of concepts found, and only the theory of numbers, perhaps, contains methods of proof of comparable beauty. It is no wonder, then, that everyone who studies the theory of sets is indescribably fascinated by it. Over and above that, however, this theory has become of the very greatest importance for the whole of mathematics. It has enriched nearly every part of mathematics, and lent it a new appearance. It has given rise to new branches of mathematics, or at least first rendered possible their further development, such as the theory of sets of points, the theory of real functions, and topology. Finally, the theory of sets has had particular influence on the investigation of the foundations of mathematics, acting in this respect, as well as through the generality of its concepts, as a connecting link between mathematics and philosophy.

Here we shall present the basic features of the *general theory of sets*. The theory of point sets will be merely touched upon.

CHAPTER I

The Rudiments of Set Theory

1. A First Classification of Sets

What is a set? By a set $\mathfrak{M}$ we are to understand, according to G. Cantor, "a collection into a whole, of definite, well-distinguished objects (called the 'elements' of $\mathfrak{M}$) of our perception or of our thought."

For example, the prime numbers between 1 and 100 constitute a set of 25 elements; the totality of even numbers, a set of infinitely many elements; the vertices of a square, a set of 4 elements; the points of a circle, a set of infinitely many elements.

For a set, the order of succession of its elements shall not matter, provided that nothing is said to the contrary. Thus, e.g., the set[1] $\{1, 2, 3\}$, consisting of the elements 1, 2, 3, is the same set as $\{3, 1, 2\}$ or $\{2, 3, 1\}$. Furthermore, the same element shall not be allowed to appear more than once. The number complex 1, 2, 1, 2, 3, consequently, becomes a set only after deleting the repeated elements.

Two sets $\mathfrak{M}$ and $\mathfrak{N}$ are said to be equal, in symbols: $\mathfrak{M} = \mathfrak{N}$, if they contain the same elements; i. e., if every element of $\mathfrak{M}$ is also an element of $\mathfrak{N}$, and, conversely, every element of $\mathfrak{N}$ also belongs to $\mathfrak{M}$. For example, $\{1, 2, 3\} = \{3, 1, 2\}$. $\mathfrak{M} \neq \mathfrak{N}$ shall mean that $\mathfrak{M}$ is not equal to $\mathfrak{N}$. If m is an element of $\mathfrak{M}$, we write $m \in \mathfrak{M}$ (read: m is an element of $\mathfrak{M}$), whereas $m \notin \mathfrak{M}$ shall denote that m is not an element of $\mathfrak{M}$.

A first, coarse classification of sets distinguishes them into finite and infinite sets, according as the sets do, or do not, contain a finite number of elements. From the infinite sets, the set of natural numbers, whose elements we may think of as being given in their natural order of succession $\{1, 2, 3, \cdots\}$,

[1]Sets are frequently designated by enclosing their elements in braces.

is especially singled out, and is called an enumerable set. More generally, an infinite set $\mathfrak{M}$ is said to be enumerable if, and only if, it can be be written as a sequence $\{m_1, m_2, m_3, \cdots\}$; i. e., if, and only if, to every element m of the set, a natural number can be made to correspond in such a manner, that to every element of the set corresponds precisely one natural number, and to every natural number corresponds precisely one element of the set.

The letters in all the printing presses on the earth form a finite set, although the number of elements is "very large." The same is true of the number of volumes in the "universal library" of K. Lasswitz's,[2] which is so large, that the librarian, were he to dash along the row of books even with the velocity of light, would arrive at the last volume only after $10^{1,999,982}$ years.

If a set is finite or enumerable, we shall say that it is at most enumerable. If it is neither finite nor enumerable, it will be called nonenumerable.

2. *Three Remarkable Examples of Enumerable Sets*

G. Cantor proved, already in one of his first papers on set theory, the enumerability of two sets which hardly seem at first glance to possess this property.

THEOREM 1. *The set of all rational numbers is enumerable.*

Proof: Let us first deal only with the positive rational numbers. We can imagine to be written down in order of magnitude, first, all whole numbers, i. e., all numbers with denominator 1; then, all fractions with denominator 2; then, all fractions with denominator 3; etc. There arise in this manner the rows of numbers

[2]See his book, "Traumkristalle".

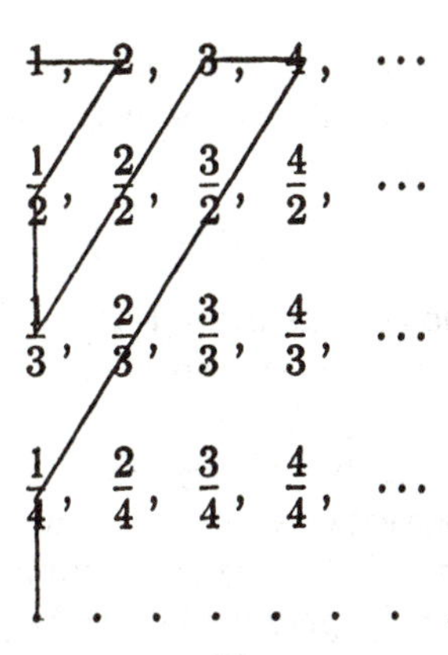

If we write down the numbers in the order of succession indicated by the line drawn in (leaving out numbers which have already appeared), then every positive rational number certainly appears, and also only once. The totality of these rational numbers is thus written as a sequence

$$1, 2, \tfrac{1}{2}, \tfrac{1}{3}, 3, 4, \tfrac{3}{2}, \tfrac{2}{3}, \tfrac{1}{4}, \cdots .$$

If we denote this sequence by $\{r_1, r_2, r_3, \cdots\}$, then obviously $\{0, -r_1, r_1, -r_2, r_2, \cdots\}$ is the set of all rational numbers, and the enumerability of this set is herewith established.

For Cantor's second theorem, which asserts the enumerability of an even "more extensive" class of numbers, let us recall the definition of algebraic number. A number of this sort is understood to be one which is a root of a polynomial

$$f(x) = a_n x^n + a_{n-1}x^{n-1} + \cdots + a_1 x + a_0 ,$$

where $a_n \neq 0$ and all a_k's are integral rational numbers. Algebraic numbers include, among others, all rational numbers and all roots of such.

THEOREM 2. *The set of all algebraic numbers is enumerable.*

Proof:[3] Let $f(x)$ be a polynomial of the kind just described:

---

[3]We confine ourselves here to *real* algebraic numbers. The theorem and proof are also valid, however, for complex algebraic numbers.

We may suppose, moreover, without loss of generality, that $a_n > 0$. Define the "height" of the polynomial as the positive number

$$h = n + a_n + |a_{n-1}| + \cdots + |a_1| + |a_0|.$$

The height is obviously an integer ≥ 1. The same height is possessed by only a finite number of polynomials, because $n \leq h$ and every $|a_k| \leq h$. Therefore to every height correspond only a finite number of algebraic numbers. This makes it possible to write the set of all algebraic numbers as a sequence. First we write down all the algebraic numbers yielded by the height 2. Since the only polynomials of height 2 are x and 2, we obtain the sole number 0. The polynomials of height 3 are x^2, $2x$, $x + 1$, $x - 1$, 3. These give the new roots -1 and $+1$. The new roots arising from the polynomials of height 4 are, in order of magnitude: -2, $-\frac{1}{2}$, $+\frac{1}{2}$, $+2$. Height 5 yields -3, $-\frac{1}{2} - \frac{1}{2}\sqrt{5}$, $-\sqrt{2}$, $-\frac{1}{2}\sqrt{2}$, $\frac{1}{2} - \frac{1}{2}\sqrt{5}$, $-\frac{1}{3}$, $\frac{1}{3}$, $-\frac{1}{2} + \frac{1}{2}\sqrt{5}$, $\frac{1}{2}\sqrt{2}$, $\sqrt{2}$, $\frac{1}{2} + \frac{1}{2}\sqrt{5}$, 3; etc. Thus, by allowing the height to run through the series of natural numbers, and writing down the newly arising, finitely many algebraic numbers corresponding to each value of the height, we obtain a sequence of distinct algebraic numbers. Since every polynomial has a height, all algebraic numbers appear in the sequence. This completes the proof of the theorem.

That the concept of enumerable set also leads to valuable results concerning functions is shown by

THEOREM 3. *Every function $f(x)$ which is monotonic in an interval $a \leq x \leq b$ is discontinuous at an at most enumerable number of points of this interval.*

Proof: It suffices to carry out the proof for monotonically increasing functions. Let $f(x)$ be such a function. It is discontinuous at a point ξ if, and only if,[4]

$$\sigma(\xi) = f(\xi + 0) - f(\xi - 0) > 0,$$

---

[4] As usual, let $f(\xi - 0)$ denote the left-hand limit and $f(\xi + 0)$ the right-hand limit of $f(x)$ at the point ξ.

where we agree that $f(a - 0) = f(a), f(b + 0) = f(b)$. If

$$a < \xi_1 < \xi_2 < \cdots < \xi_p < b,$$

and $x_1, \cdots, x_{p-1}$ are numbers in the intervals

$$\xi_\nu < x_\nu < \xi_{\nu+1},$$

and if, finally, $x_0 = a$, $x_p = b$, then

$$f(x_\nu) - f(x_{\nu-1}) \geq f(\xi_\nu + 0) - f(\xi_\nu - 0) = \sigma(\xi_\nu),$$

so that

$$f(b) - f(a) = \sum_{\nu=1}^{p} (f(x_\nu) - f(x_{\nu-1})) \geq \sum_{\nu=1}^{p} \sigma(\xi_\nu).$$

If, now, the ξ_ν are numbers with $\sigma(\xi_\nu) > 1/n$, it follows that

$$p < n(f(b) - f(a));$$

i. e., the number of points of discontinuity ξ with $\sigma(\xi) > 1/n$ has a fixed upper bound. Hence, at most a finite number of points of discontinuity ξ with $\sigma(\xi) > 1/n$ belong to the interval $a \leq x \leq b$, even when the end points have been taken into account. The totality of points of discontinuity can therefore be written as a sequence, by first writing down the finite number of points of discontinuity ξ with $\sigma(\xi) > 1$, next the newly arising ones with $\sigma(\xi) > \frac{1}{2}$, then the newly acquired ones with $\sigma(\xi) > \frac{1}{3}$, etc. Since to each point of discontinuity ξ there belongs a definite positive number $\sigma(\xi)$, every point of discontinuity appears in this sequence, and indeed precisely once. Thus our assertion is proved.

3. *Subset, Sum, and Intersection of Sets, in Particular, of Enumerable Sets*

The considerations which led in §2 to the enumerability of the set of all rational numbers, furnish, over and above this, a general theorem which is often useful for establishing the enumerability of a set. Before we formulate this theorem, we shall introduce several new concepts which will be encountered continually in what follows.

Let a set $\mathfrak{M}$ be given. The set $\mathfrak{N}$ is called a subset or part of $\mathfrak{M}$, in symbols: $\mathfrak{N} \subseteq \mathfrak{M}$, if every element of $\mathfrak{N}$ is at the same time an element of $\mathfrak{M}$; i. e., if $n \in \mathfrak{N}$ always implies $n \in \mathfrak{M}$. E. g., according to this, every set is a subset of itself, and, in fact, it is called an improper subset. On the other hand, $\mathfrak{N}$ is called a proper subset of $\mathfrak{M}$, in symbols: $\mathfrak{N} \subset \mathfrak{M}$, if $\mathfrak{N}$ is a subset of $\mathfrak{M}$ and $\mathfrak{N} \neq \mathfrak{M}$. The set of those elements of $\mathfrak{M}$ which do not belong at the same time to the subset $\mathfrak{N}$, is termed the residual set $\mathfrak{R}$ belonging to $\mathfrak{N}$, or the complement of $\mathfrak{N}$ with respect to $\mathfrak{M}$; in symbols: $\mathfrak{R} = \mathfrak{M} - \mathfrak{N}$. So that this definition may be valid also for the case in which $\mathfrak{N}$ is an improper subset of $\mathfrak{M}$, we introduce an ideal set, the so-called null set or empty set, which we denote by 0.[5] This is analogous to the introduction, in geometry, of infinitely distant points as ideal points, and the introduction, at one time, of zero into the number system, as an ideal number.[6] The null set shall be classed with the finite sets. It shall be a subset of every set, and, in particular, of itself.

Examples. The boundary points $x^2 + y^2 = 1$ form a proper subset of the points of the circle $x^2 + y^2 \leq 1$; its complement consists of the interior points $x^2 + y^2 < 1$ of the circle. The set of rational numbers is a subset of the set of all real numbers; the complementary set consists of the irrational numbers.

Let there be given an enumerable set $\mathfrak{M} = \{m_1, m_2, m_3, \cdots\}$, and suppose that $0 \subset \mathfrak{N} \subseteq \mathfrak{M}$. Then, in the sequence $\mathfrak{M}$, there is a first element m_{k_1} which belongs to $\mathfrak{N}$; let $n_1 = m_{k_1}$. This is followed again by a first element, m_{k_2}, in $\mathfrak{M}$, which belongs to $\mathfrak{N}$; let $n_2 = m_{k_2}$; etc. The process does or does not terminate, according as $\mathfrak{N}$ is finite or not. Since $\mathfrak{M}$ contains all the elements of $\mathfrak{N}$ (and perhaps even more), the possibly terminating sequence $\{n_1, n_2, n_3, \cdots\}$ comprises precisely the elements of $\mathfrak{N}$. Therefore $\mathfrak{N}$ is finite or enumerable. Hence, we have

---

[5]One must distinguish the empty set 0 from the set $\{0\}$ containing the number 0 as an element.

[6]Here we are thinking of the historical development of the number concept, not, say, of the role which the number 0 plays in an axiomatic construction of the number system.

THEOREM 1. *Every subset of an enumerable set is at most enumerable.*

E. g., according to this theorem, the set of all algebraic numbers which satisfy irreducible equations of prime degree is enumerable as a part of the set of all algebraic numbers. The set of positive fractions which can be represented as the sum of three biquadrates of rational numbers is enumerable as a subset of the set of all rational numbers.

The union or (logical) sum $\mathfrak{S}$ of finitely or infinitely many sets is understood to be the set of those elements which belong to *at least* one of the sets. The union of an at most enumerable number of sets $\mathfrak{M}_1$, $\mathfrak{M}_2$, $\mathfrak{M}_3$, $\cdots$ is written in the form

$$\mathfrak{S} = \mathfrak{M}_1 + \mathfrak{M}_2 + \cdots \qquad \text{or} \qquad \mathfrak{S} = \sum_{k=1}^{\infty} \mathfrak{M}_k .$$

The intersection, logical product, or product of the first kind, of arbitrarily many sets, is understood to be the set of those elements which belong to *each* of the aforesaid sets. For the intersection, $\mathfrak{D}$, of two sets $\mathfrak{M}_1$ and $\mathfrak{M}_2$, we write

$$\mathfrak{D} = \mathfrak{D}(\mathfrak{M}_1, \mathfrak{M}_2) \qquad \text{or} \qquad \mathfrak{D} = \mathfrak{M}_1 \cdot \mathfrak{M}_2 ;$$

and for the intersection of at most enumerably many sets,

$$\mathfrak{D} = \mathfrak{D}(\mathfrak{M}_1, \mathfrak{M}_2, \cdots) \qquad \text{or} \qquad \mathfrak{D} = \mathfrak{M}_1 \cdot \mathfrak{M}_2 \cdots = \prod_{k=1}^{\infty} \mathfrak{M}_k .$$

The second notation, which will be employed here by preference, offers the advantage that rules of operation for the sum and product of sets then hold, which are similar to those for the sum and product of numbers. For example,

$$\mathfrak{M}(\mathfrak{P} + \mathfrak{Q}) = \mathfrak{M} \cdot \mathfrak{P} + \mathfrak{M} \cdot \mathfrak{Q}.$$

In forming sums, null sets are left out, and the sum of null sets shall be the null set. If, in a product, one of the sets is empty, we put $\mathfrak{D} = 0$; likewise if the sets have no element in common. If the intersection of every pair of sets under consideration is 0, the sets are called mutually exclusive or disjunct.

Examples. If $\mathfrak{M} = \{1, 2, 3, \cdots\}$ and $\mathfrak{N} = \{5, 7, 9\}$, then $\mathfrak{M} + \mathfrak{N} = \mathfrak{M}$ and $\mathfrak{M} \cdot \mathfrak{N} = \mathfrak{N}$. If $\mathfrak{M}$ is the set of even integers and $\mathfrak{N}$ is the set of odd integers, then $\mathfrak{M} + \mathfrak{N}$ is the set of all integers and $\mathfrak{M} \cdot \mathfrak{N} = 0$. The two circles in Fig. 1 have as sum the entire hatched region, and as product, the crosshatched region. The enumerably many sets $\{1, 2, 3, 4, \cdots\}$, $\{2, 3, 4, \cdots\}$, $\{3, 4, \cdots\}$, $\cdots$ have the first set as their sum, and their intersection is 0.

The theorem alluded to at the beginning of this §3 now reads as follows:

THEOREM 2. *The sum of at most enumerably many sets, each of which is at most enumerable, is likewise at most enumerable.*

Proof: If we are dealing with two sets, $\mathfrak{M} = \{m_1, m_2, \cdots\}$, $\mathfrak{N} = \{n_1, n_2, \cdots\}$, which are finite or enumerable, we can form the sequence $m_1, n_1, m_2, n_2, \cdots$ so far as the given sets furnish elements as contributions. If we strike out in this sequence every element that coincides with a preceding one, we have obviously written the set $\mathfrak{M} + \mathfrak{N}$ as a sequence and proved the enumerability of the sum. In this way the proof for finitely many sets can always be carried out by "sliding together" the given sets.

If there are, however, enumerably many sets, $\mathfrak{M}_1, \mathfrak{M}_2, \cdots$, let $\mathfrak{M}_\mu$ have the elements $m_{\mu 1}, m_{\mu 2}, \cdots$. We first write down, one under another, the sets given as sequences:

$$
\begin{array}{llllr}
m_{11}, & m_{12}, & m_{13}, & \cdots & (\mathfrak{M}_1) \\
m_{21}, & m_{22}, & m_{23}, & \cdots & (\mathfrak{M}_2) \\
m_{31}, & m_{32}, & m_{33}, & \cdots & (\mathfrak{M}_3) \\
\cdot & \cdot & \cdot & \cdots & ,
\end{array}
$$

and then write down the elements anew in the order of succession indicated by the arrows:

$$m_{11}, m_{21}, m_{12}, m_{31}, m_{22}, m_{13}, \cdots,$$

leaving out elements that have already appeared. We thus get a sequence containing all the elements that occur in any of our $\mathfrak{M}_\mu$'s, thereby proving the assertion.

The principle of this proof, which is obviously that already employed on p. 3 and is also involved in the proof on p. 4 is called the first (or Cauchy) diagonal-method.

Examples. In the plane, the set of all points whose coordinates are both rational numbers is enumerable. For if the rational abscissa x, say, is held fixed, the points with this fixed abscissa and arbitrary rational ordinate y form, according to §2, Theorem

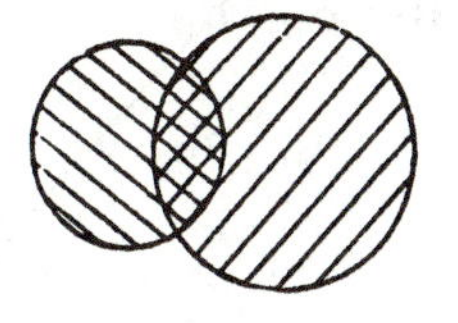

Fig. 1.

1, an enumerable set. If we now permit x also to run through all the rational numbers, we have enumerably many enumerable sets, which thus all together yield once more an enumerable set. It follows, likewise, in space, that the set of all points whose coordinates are all rational numbers is enumerable. In particular, it is a consequence of this, together with Theorem 1, that the set of all cubes with sides of length 1 and vertices with integral coordinates is enumerable. Further, in the plane, the set of all circles with rational centers and radii is enumerable.

4. *An Example of a Nonenumerable Set*

According to what has preceded, there is still the possibility of all infinite sets being enumerable. The distinction of sets into enumerable and nonenumerable ones becomes significant only after the existence of nonenumerable sets has been demonstrated. We therefore prove the following

THEOREM. *The set of all real numbers in the interval* $0 \leq x \leq 1$ *is nonenumerable.*

Proof: The proof is also carried out by means of a so-called diagonal method; let us call it the second (or Cantor) diagonal-method. On account of §3, Theorem 1, it is sufficient to demonstrate the nonenumerability of the set $0 < x \leq 1$. Every number $0 < x \leq 1$ can be written as a nonterminating decimal fraction $0 \cdot a_1 a_2 a_3 \cdots$ (e. g., $\frac{1}{2} = 0.499 \cdots$, $1 = 0.999 \cdots$), and, in fact, in a unique manner. If, now, the numbers in the interval $0 < x \leq 1$ were enumerable, they could be written as a sequence of nonterminating decimal fractions

$$(1) \qquad \begin{cases} 0 \cdot & a_{11} & a_{12} & a_{13} & \cdots \\ 0 \cdot & a_{21} & a_{22} & a_{23} & \cdots \\ 0 \cdot & a_{31} & a_{32} & a_{33} & \cdots \\ \cdot & \cdot & \cdot & \cdot & \cdot \end{cases}$$

From the principal diagonal indicated by the line, we now form the nonterminating decimal fraction $0 \cdot a_{11} a_{22} a_{33} \cdots$, and from this one we construct a new decimal fraction by replacing each digit a_{nn} by a different one, b_n, where the latter is taken different from 0. We have then for the resulting decimal fraction, $d = 0 \cdot b_1 b_2 b_3 \cdots$, also $0 < d \leq 1$. The decimal fraction for d does not terminate because of the absence of zeros. It should therefore have to coincide in all digits with one of the decimal fractions in (1). That, however, according to our construction of d, is certainly not the case, because d surely differs from each nth decimal fraction in (1) in the nth digit. This completes the proof of the theorem.

From this theorem we obtain the following very remarkable

Corollary. The set of all transcendental numbers[7] *is nonenumerable.*

For if the set were enumerable, the addition of the set of algebraic numbers, which, according to §2, Theorem 2, is enumerable, would again result in only an enumerable set. This

[7]A number is called transcendental, if it is not an algebraic number. Hermite, in 1873, proved that the number e is transcendental, and Lindemann, in 1892, established the transcendentality of the number π.

means that the set of all real numbers, and hence, *a fortiori*, the subset of real numbers lying in the interval $0 < x \leq 1$, would be enumerable. This would, however, contradict the theorem just proved.

CHAPTER II

Arbitrary Sets and Their Cardinal Numbers

1. Extensions of the Number Concept

Up to now we have classified sets into finite, enumerable, and nonenumerable sets. To the last class belong simply all those sets that remain after the first two classes have been split off. One can raise the question, whether it is not possible to further subdivide the class of nonenumerable sets. G. Cantor gave this question, in truly ingenious fashion, the following turn: Can the concept of natural number be generalized in such a manner, that every set is assigned one of these generalized "numbers" for the "number of its elements", as it were? Should this be possible, there would result immediately a classification of infinite sets, too, according to the "number of their elements." Clearly these new "numbers" would have to be something quite novel. It is therefore useful to recall how it is otherwise customary in mathematics to introduce new numbers.

The number concept has, of course, been extended several times in the evolution of mathematics. The first extension consisted in the introduction of fractions. In a rigorous introduction of the rational numbers, one must, by all means, abandon the primitive method of "dividing a whole into a certain number of parts", and instead of this, proceed as follows: One considers, as a new kind of "number", a pair of natural numbers, a, b, which, with this interpretation, shall be written in the form a/b, in conformity with the usual manner of writing fractions. This, however, would not yet accomplish altogether what one would like to have. For, the number pairs a/b, $2a/2b$, $3a/3b$ are all to be regarded merely as different representations of the same rational number. The correct introduction of the rational number can therefore occur as follows: In the first place, one makes the stipulation that every pair of number pairs pa/pb and qa/qb, composed of natural numbers, shall be regarded as equivalent. In the second place, one agrees that a rational number shall be understood to be an arbitrary representative chosen from a class consisting of equivalent number pairs. In deriving the rules of operation for rational numbers, one must always make sure that these rules of operation are independent of which representative of the rational number is chosen in any given instance.

Nowadays the irrational number is usually introduced by means of nests of intervals. A nest of intervals is a sequence of closed intervals

$$\langle r_1, s_1\rangle, \quad \langle r_2, s_2\rangle, \quad \langle r_3, s_3\rangle, \cdots,$$

where the r_n's and s_n's are rational numbers such that always

$$r_{n-1} \leq r_n \leq s_n \leq s_{n-1}$$

and $s_n - r_n \to 0$ as $n \to \infty$. Two nests of intervals, consisting of the interval sequences $\langle r_n, s_n\rangle$ and $\langle \bar{r}_n, \bar{s}_n\rangle$, are regarded as "equivalent," if $\bar{r}_n \leq s_n$ and $r_n \leq \bar{s}_n$ for all n. Analogously to what was stated above, an irrational number or, in general, a real number is an arbitrary representative taken from a class composed of equivalent nests of intervals. Here, too, in deriving the rules of operation for real numbers, one must make certain that they are independent of which representative of the real number is chosen in any particular instance.

Let us bear these things in mind for the introduction of cardinal numbers.

2. *Equivalence of Sets*

A second point of departure for Cantor's extension of the number concept is, as opposed to the first, of a very simple nature. A child who is unable to count can nevertheless determine whether there are, for example, just as many chairs as persons present in a room. He need only have each person take a seat on a chair. By this act, pairs are formed, each pair consisting of one person and one chair. One can also say that the chairs and persons are made to correspond to each other in a one-to-one manner; i. e., in such a way that precisely one chair corresponds to each person, and exactly one person corresponds to each chair. This primitive procedure, however, can be carried over to arbitrary sets too, and it then leads to a concept which corresponds to that of the "same number" of elements in the case of finite sets. We accordingly set down the following

Definition. A set $\mathfrak{M}$ *is said to be equivalent to a set* $\mathfrak{N}$*, in symbols:* $\mathfrak{M} \sim \mathfrak{N}$*, if it is possible to make the elements of* $\mathfrak{N}$ *correspond to the elements of* $\mathfrak{M}$ *in a one-to-one manner; i. e., if it is possible to make correspond to every element* m *of* $\mathfrak{M}$ *an element* n *of* $\mathfrak{N}$ *in such a manner that, on the basis of this correspondence, to every element of* $\mathfrak{M}$ *there corresponds one, and only one, element of* $\mathfrak{N}$*, and, conversely, to every element of* $\mathfrak{N}$*, one, and only one, element of* $\mathfrak{M}$*. Instead of saying one-to-one correspondence, we also speak, for brevity, of a mapping. The empty set shall be equivalent only to itself.*

As with all definitions of this kind (equality, similarity, etc.), it is to be required of the concept of equivalence, that it be reflexive, symmetric, and transitive; i. e., that it possess the following properties:

α) $\mathfrak{M} \sim \mathfrak{M}$; i. e., every set is equivalent to itself;

β) $\mathfrak{M} \sim \mathfrak{N}$ implies $\mathfrak{N} \sim \mathfrak{M}$;

γ) if $\mathfrak{M} \sim \mathfrak{N}$ and $\mathfrak{N} \sim \mathfrak{P}$, then $\mathfrak{M} \sim \mathfrak{P}$.

These three laws indeed follow immediately from the definition.

Since the concept of equivalence is of really fundamental importance, we shall illustrate it by a series of examples to which we shall also occasionally refer later on.

a) The sets of points of the intervals[1] $\langle 0, 1\rangle$, $\langle 0, 1)$, $(0, 1\rangle$, $(0, 1)$ are equivalent to each other.

First let us prove that $(0, 1\rangle \sim (0, 1)$. We denote the points of the first of the two intervals by x, and those of the second, by y, and set up the following correspondence:

[1] $\langle a, b\rangle$ denotes a *closed* interval; (a, b), an *open* interval; $\langle a, b)$ and $(a, b\rangle$, *half-open* intervals. These are intervals in which both end points belong, no end point belongs, the left-hand end point belongs, the right-hand end point belongs, to the respective interval.

$$y = \tfrac{3}{2} - x \quad \text{for } \tfrac{1}{2} < x \leq 1; \quad \text{then we have } \tfrac{1}{2} \leq y < 1;$$

$$y = \tfrac{3}{4} - x \quad \text{for } \tfrac{1}{4} < x \leq \tfrac{1}{2}; \quad \text{then we have } \tfrac{1}{4} \leq y < \tfrac{1}{2};$$

$$y = \tfrac{3}{8} - x \quad \text{for } \tfrac{1}{8} < x \leq \tfrac{1}{4}; \quad \text{then we have } \tfrac{1}{8} \leq y < \tfrac{1}{4};$$

etc.

It is evident that to every x of the first interval there is hereby made to correspond one, and only one, point y of the second interval, and, conversely, to every y, one, and only one, x. This proves our assertion.

One shows in an analogous fashion, that $\langle 0, 1) \sim (0, 1)$. From this it follows finally, however, that also $\langle 0, 1) \sim \langle 0, 1\rangle$. For we can associate the point 0 of the first interval with the point 0 of the second, and set up a one-to-one correspondence between the remaining points of the two intervals n the manner prescribed in the preceding paragraph.

b) For the points of any two finite intervals $\mathfrak{J}_1$ and $\mathfrak{J}_2$ we have invariably $\mathfrak{J}_1 \sim \mathfrak{J}_2$.

It suffices to show that one of these two intervals, say $\mathfrak{J}_1$, is equivalent to one of the intervals mentioned in a). Choose this interval so, that if it be applied to a line parallel to $\mathfrak{J}_1$, it be circumstanced exactly as $\mathfrak{J}_1$ with regard to the appurtenance of its limit points to the interval. Then it is easy to bring about a one-to-one correspondence between the points of the two intervals by means of a central projection, as indicated in Figure 2.

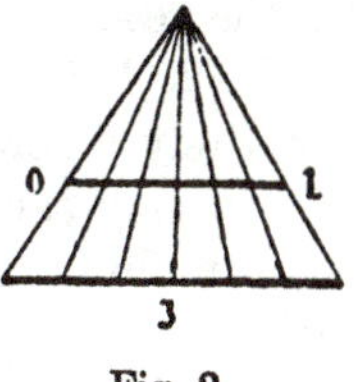

Fig. 2.

c) A half-line and an entire line are equivalent to an interval. For, a half-line can be obtained by central projection, from an interval erected at right angles to it and open above (Fig. 3); and a full line, in a similar manner, from a bent, open interval (Fig. 4).

d) The interval $\langle 0, 1\rangle$ is often designated as the continuum. The results established above then state that all intervals, half-lines, and complete straight lines are equivalent to one another and, in particular, to the continuum.

e) Two finite sets are obviously equivalent if, and only if, they contain equally many elements.

f) All enumerable sets are equivalent to each other, but not to any finite set.

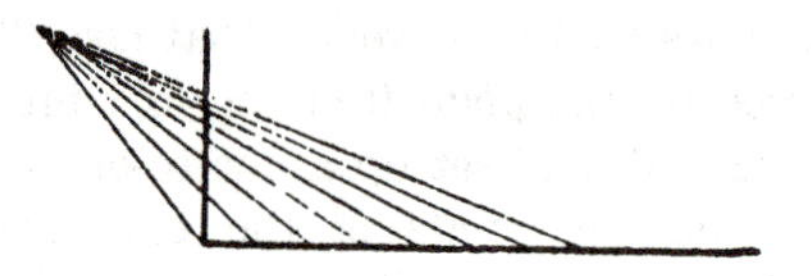

Fig. 3.

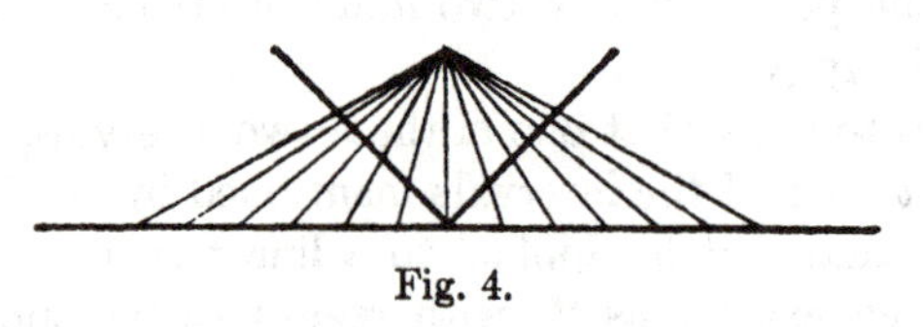

Fig. 4.

The first follows from the fact that every enumerable set is, according to its definition (p. 2), equivalent to the set $\{1, 2, 3, \cdots\}$. The second is obvious.

g) An infinite set can be equivalent to one of its proper subsets. This is shown by the result d) or the equivalence of the sets $\{1, 2, 3, \cdots\}$ and $\{2, 4, 6, \cdots\}$. Infinite sets thus exhibit in this respect an entirely different behavior from that of finite sets (cf. e)). This property can therefore also be used to fix the distinction between finite and infinite sets, independently of any enumeration.

h) There exist infinite sets which are not equivalent to each other; namely, as follows from ch. I, §4, the continuum on the one hand and the enumerable sets on the other. It is conceivable, however, that all nonenumerable sets might be equivalent to one another. In that case, the introduction of the concept of equivalence would bring about no new classification of sets. The following result is therefore of fundamental importance:

i) There exist infinite sets which are neither enumerable nor equivalent to the continuum. An example of such a set is the set of all real functions defined in the interval $\langle 0, 1 \rangle$.

The proof of this is obtained by a procedure which is again essentially the second diagonal method (cf. p. 9). The concept of function which is taken here as basis is Dirichlet's general concept of a function. According to this, a function $y = f(x)$ is defined in an interval, if, with every x of the interval, there is associated a well-determined number y. Further, two functions are to be designated as distinct, even if they possess different functional values at merely a single point of the interval. Now it is clear not only that there are infinitely many distinct functions in the interval $\langle 0, 1 \rangle$, but also that there exist nonenumerably many. For, at the point $x = 0$ alone, the functions can already assume the nonenumerably many values between 0 and 1. Thus, there remains to be proved only that the set of functions is not equivalent to the continuum. Suppose, now, that the set of functions were equivalent to the continuum. Then it would be possible to make the functions $f(x)$ and the points z of the interval $\langle 0, 1 \rangle$ correspond in a one-to-one manner. Denote by $f_z(x)$ the function thus assigned to the point z. Now construct a function $g(z)$ in the interval $0 \leq z \leq 1$, with the property that, at every point z, $g(z) \neq f_z(z)$. (This can be accomplished in various ways.) Since $g(z)$ is also a function defined in the interval $\langle 0, 1 \rangle$, it must coincide with an $f_u(x)$; hence, in particular, $g(u) = f_u(u)$. This, however, is excluded by the definition of $g(z)$. The assumption that the set of our functions is equivalent to the continuum has thus led to a contradiction.

3. *Cardinal Numbers*

The introduction of the heralded "numbers" offers no difficulty now after the preliminaries encountered in §1 and §2.

Corresponding exactly to the definition of the rational number given on p. 13, we set down the following

Definition 1. *By a cardinal number of a power* $\mathfrak{m}$ *we mean an arbitrary representative* $\mathfrak{M}$ *of a class of mutually equivalent sets. The cardinal number or power of a set* $\mathfrak{M}$ *will also be denoted by* $|\mathfrak{M}|$. *The notation* $|\mathfrak{M}|$ *accordingly means simply that the set* $\mathfrak{M}$ *may be replaced by any set equivalent to it.*

Abstract as this definition may sound, it does nevertheless accord precisely with the grammatical sense of the words "cardinal number" in the case of *finite* sets. For, the number of elements—and the cardinal numbers of the language serve to designate this number—is determined, for finite sets, by counting; e. g., the number 4, by counting 1, 2, 3, 4. That is, the number 4 is represented by the set $\{1, 2, 3, 4\}$. In virtue of §2, e, this coincides exactly with the definition of cardinal number given above. Consequently, for a finite set, the cardinal number introduced above is precisely the number, n, of its elements, and shall also be denoted simply by n.

Of infinite sets we know as yet but a few classes of equivalent sets, namely: the enumerable sets, which may be represented by the set $\{1, 2, 3, \cdots\}$, say; further, those sets which are equivalent to the continuum; and those that are equivalent to the set of functions in §2, i. Let us introduce the abbreviations $\mathfrak{a}$, $\mathfrak{c}$, $\mathfrak{f}$ for the cardinal numbers of these sets. If, finally, we assign the cardinal number 0 to the empty set, we are familiar thus far with the cardinal numbers

$$0, 1, 2, 3, \cdots; \quad \mathfrak{a}, \mathfrak{c}, \mathfrak{f}.$$

It is necessary now to decide when one of two cardinal numbers shall be regarded as the smaller of the two. This is accomplished by

Definition 2. *A set* $\mathfrak{M}$ *is said to have a smaller cardinal number, or a lesser power, than a set* $\mathfrak{N}$, *in symbols*: $|\mathfrak{M}| < |\mathfrak{N}|$, *if, and only if,* $\mathfrak{M}$ *is equivalent to a subset of* $\mathfrak{N}$, *but* $\mathfrak{N}$ *is equivalent to no subset of* $\mathfrak{M}$.

This definition makes an agreement concerning the cardinal

numbers of two sets, on the basis of certain relations between the sets themselves. Therein lies the assumption that these relations are preserved if the sets in them are replaced by equivalent sets. Now, for the definition to have a meaning, this must first be demonstrated. To this end, let $\mathfrak{M} \sim \overline{\mathfrak{M}}$ and $\mathfrak{N} \sim \overline{\mathfrak{N}}$. Further, suppose that $\mathfrak{M} \sim \mathfrak{N}_1$ for a subset $\mathfrak{N}_1$ of $\mathfrak{N}$, but that $\mathfrak{N}$ is equivalent to no subset of $\mathfrak{M}$. Since $\mathfrak{N} \sim \overline{\mathfrak{N}}$, there is a one-to-one correspondence between the elements of $\mathfrak{N}$ and those of $\overline{\mathfrak{N}}$. A subset $\overline{\mathfrak{N}}_1$ of $\overline{\mathfrak{N}}$ is thereby made to correspond to the elements of $\mathfrak{N}_1$. Then $\overline{\mathfrak{N}}_1 \sim \mathfrak{N}_1 \sim \mathfrak{M} \sim \overline{\mathfrak{M}}$, and hence $\overline{\mathfrak{N}}_1 \sim \overline{\mathfrak{M}}$. Now we have to show merely that $\overline{\mathfrak{N}}$ is equivalent to no subset of $\overline{\mathfrak{M}}$. Were this not the case, the procedure just employed would prove $\mathfrak{N}$ to be equivalent to a subset of $\mathfrak{M}$, contradicting the hypothesis. The justification of the definition is thus established.

It is clear that the above definition, for finite cardinal numbers, yields the customary "less than" concept.

Beyond this, however, it is desirable that the familiar laws governing inequalities involving finite numbers hold also for cardinal numbers. We refer to the following three rules:

α) If $\mathfrak{m} < \mathfrak{n}$ and $\mathfrak{n} < \mathfrak{p}$, then $\mathfrak{m} < \mathfrak{p}$.

β) If $\mathfrak{m}, \mathfrak{n}$ are any two cardinal numbers, then *at most* one of the three relations $\mathfrak{m} < \mathfrak{n}$, $\mathfrak{m} = \mathfrak{n}$, $\mathfrak{n} < \mathfrak{m}$ holds; and hence, in particular, if $\mathfrak{m} \leq \mathfrak{n}$ and $\mathfrak{n} \leq \mathfrak{m}$, then invariably $\mathfrak{m} = \mathfrak{n}$.

γ) If $\mathfrak{m}, \mathfrak{n}$ are any two cardinal numbers, then *at least* one of the three relations $\mathfrak{m} < \mathfrak{n}$, $\mathfrak{m} = \mathfrak{n}$, $\mathfrak{n} < \mathfrak{m}$ subsists.

If both β) and γ) are true, it follows that *precisely one* of these three relations holds.

To α): Let $\mathfrak{M}, \mathfrak{N}, \mathfrak{P}$ be representatives of the cardinal numbers $\mathfrak{m}, \mathfrak{n}, \mathfrak{p}$. Then, by hypothesis, there exist sets $\mathfrak{N}_1$ and $\mathfrak{P}_1$ such that

$$\mathfrak{N}_1 \subseteq \mathfrak{N}, \quad \mathfrak{P}_1 \subseteq \mathfrak{P}, \quad \mathfrak{M} \sim \mathfrak{N}_1, \quad \mathfrak{N} \sim \mathfrak{P}_1 .$$

Because of the last equivalence, there is a mapping of the set $\mathfrak{N}$ on $\mathfrak{P}_1$. By means of this mapping, $\mathfrak{N}_1$ in particular is mapped on a subset $\mathfrak{P}_2$ of $\mathfrak{P}_1$. Then $\mathfrak{N}_1 \sim \mathfrak{P}_2$, and hence also $\mathfrak{M} \sim \mathfrak{P}_2$.

Now we have merely to show that $\mathfrak{P}$ is equivalent to no subset of $\mathfrak{M}$. If we had say $\mathfrak{P} \sim \mathfrak{M}_1 \subseteq \mathfrak{M}$, however, then, because of $\mathfrak{M} \sim \mathfrak{N}_1$, there would exist a set $\mathfrak{N}_2 \subseteq \mathfrak{N}_1$ such that $\mathfrak{M}_1 \sim \mathfrak{N}_2$, and hence $\mathfrak{P} \sim \mathfrak{N}_2$, which would contradict $\mathfrak{n} < \mathfrak{p}$.

To β): According to the definition of "less than", the middle one of the relations β) cannot hold simultaneously with any one of the other two. If, however, one of the other two relations subsists, say $\mathfrak{m} < \mathfrak{n}$, and these two cardinal numbers are represented by the sets $\mathfrak{M}$, $\mathfrak{N}$, then $\mathfrak{M}$ is equivalent to a subset of $\mathfrak{N}$. But then we cannot have $\mathfrak{n} < \mathfrak{m}$.

To γ): This proposition is indeed true also, but can be proved only much later (p. 119) and with the help of the well-ordering theorem. It will therefore be necessary in the meantime, when operating with cardinal numbers, to take care that γ) is not used.

Stipulate, finally, that $\mathfrak{m} > \mathfrak{n}$ shall mean the same as $\mathfrak{n} < \mathfrak{m}$. This is sometimes convenient for calculation.

4. *Introductory Remarks Concerning the Scale of Cardinal Numbers*

If we apply Definition 2 of §3 to finite cardinal numbers, we obtain, as already remarked on p. 19, nothing new. If we proceed to the cardinal number of infinite sets, or, as we say briefly, to transfinite cardinal numbers, we can establish the following for those cardinal numbers with which we are familiar:

a) If $\mathfrak{m}$ is any transfinite cardinal number, and n is finite, then $\mathfrak{m} > n$.

For if $\mathfrak{M}$ and $\mathfrak{N}$ are representatives of $\mathfrak{m}$ and n, $\mathfrak{M}$ contains a subset equivalent to $\mathfrak{N}$, namely, every subset of n elements. On the other hand, $\mathfrak{M}$, being an infinite set, is equivalent to no subset of the finite set $\mathfrak{N}$.

b) For every transfinite cardinal number $\mathfrak{m}$, $\mathfrak{m} \geq \mathfrak{a}$. Thus the number $\mathfrak{a}$ is the *smallest transfinite cardinal number.*

This follows from the fact that every infinite set contains an enumerable subset.

c) Since $\mathfrak{a} \leq \mathfrak{c}$ and $\mathfrak{a} \neq \mathfrak{c}$, we have $\mathfrak{a} < \mathfrak{c}$. Whether there are any cardinal numbers between $\mathfrak{a}$ and $\mathfrak{c}$ is not known to this day, despite the most strenuous efforts to settle this question, which is the substance of the so-called continuum problem.[2]

d) $\mathfrak{a} < \mathfrak{c} < \mathfrak{f}$.

Because of c) we have merely to prove that $\mathfrak{c} < \mathfrak{f}$. If $\mathfrak{C}$ denotes the continuum, and $\mathfrak{F}$ the set of functions in the interval $\langle 0, 1 \rangle$, then $\mathfrak{C}$ is equivalent to a subset of $\mathfrak{F}$. For, the functions which are constant, and, in fact, equal to a value c in the interval $\langle 0, 1 \rangle$, constitute a subset of $\mathfrak{F}$, and this subset is equivalent to $\mathfrak{C}$. Hence, we need only show that $\mathfrak{F}$ is equivalent to no subset of $\mathfrak{C}$. This follows once again from the method of proof given in §2, i, but can also be inferred, at the conclusion of §5, from Bernstein's equivalence theorem.

Up to now we know only three transfinite cardinal numbers. It is therefore natural to inquire whether there are any others. The answer is: *There are infinitely many transfinite cardinal numbers.* This follows immediately from the fact that, given any transfinite cardinal number, there exists a larger one; and this fact can be proved, moreover, in the following sharper form:

THEOREM. *For every set $\mathfrak{M}$, the set $\mathfrak{U}(\mathfrak{M})$ of all its subsets has a greater cardinal number than $\mathfrak{M}$.*

For finite sets, this statement is trivial. For example, the set of all subsets of the empty set is the set $\{0\}$, i. e., a set having one element, so that it has the cardinal number 1, whereas the empty set has the cardinal number 0. For the set $\{a\}$, $\mathfrak{U}(\{a\}) = \{0, \{a\}\}$, so that the assertion here is that $1 < 2$. It is easy to see, in general, that for a set consisting of a finite number, say n, of elements, the theorem states that $n < 2^n$.

The general proof will include the case of finite sets. Let $\mathfrak{M}(m)$ denote a set with the elements m, and likewise let $\mathfrak{M}(\{m\})$ be the set which results from $\mathfrak{M}(m)$ by replacing every element m by the set $\{m\}$. Since invariably $\{m\} \subseteq \mathfrak{M}$, we have

[2] The detailed execution of an idea, due to D. Hilbert (Math. Ann. 95 (1926), 161 ff.), for a proof, still presents great difficulties.

$\mathfrak{M}(\{m\}) \subseteq \mathfrak{U}(\mathfrak{M})$ and at the same time $\mathfrak{M}(\{m\}) \sim \mathfrak{M}(m)$. According to this, $\mathfrak{M}$ is equivalent to a subset of $\mathfrak{U}(\mathfrak{M})$, and all that remains to be proved is that $\mathfrak{U}(\mathfrak{M})$ is equivalent to no subset of $\mathfrak{M}$. To this end, we shall show, in a manner after the second diagonal method, that, if $\mathfrak{U}_0 \sim \mathfrak{M}_0$ for any two sets $\mathfrak{U}_0 \subseteq \mathfrak{U}(\mathfrak{M})$ and $\mathfrak{M}_0 \subseteq \mathfrak{M}$, then $\mathfrak{U}_0 \subset \mathfrak{U}(\mathfrak{M})$. From this it then follows that $\mathfrak{U}(\mathfrak{M})$ is equivalent to no subset of $\mathfrak{M}$.

Suppose, then, that

$$\mathfrak{M}_0 \subseteq \mathfrak{M}, \quad \mathfrak{U}_0 \subseteq \mathfrak{U}(\mathfrak{M}), \quad \mathfrak{M}_0 \sim \mathfrak{U}_0 .$$

Then there exists a mapping of $\mathfrak{M}_0$ on $\mathfrak{U}_0$. Keep this mapping fixed, and let $u = \varphi(m)$ denote that element u of $\mathfrak{U}_0$ which is made to correspond, under this mapping, to the element m of $\mathfrak{M}_0$. Since every element u of $\mathfrak{U}_0$ is a subset of $\mathfrak{M}$, it is meaningful to ask whether a given element m of $\mathfrak{M}_0$ is also an element of a given u. Now let $\bar{u}$ be the set consisting of those elements m of $\mathfrak{M}_0$, each of which is not contained as an element in the corresponding $u = \varphi(m)$. This set $\bar{u}$, which is a (possibly empty) subset of $\mathfrak{M}$, does not appear as an element in $\mathfrak{U}_0$.

In fact, if $\bar{u}$ were an element of $\mathfrak{U}_0$, an element $\bar{m}$ of $\mathfrak{M}_0$ would correspond to $\bar{u}$ under the mapping fixed at the beginning. Then precisely one of the two cases $\bar{m} \in \bar{u}$ or $\bar{m} \notin \bar{u}$ would have to hold. The first case cannot occur, according to the definition of $\bar{u}$. In the second case, $\bar{m}$ would not be contained in the $\bar{u}$ corresponding to it; but then, by the definition of $\bar{u}$, $\bar{m}$ would have to belong to $\bar{u}$ all the same. The assumption $\bar{u} \in \mathfrak{U}_0$ thus invariably leads to a contradiction. Hence, $\mathfrak{U}_0 \subset \mathfrak{U}(\mathfrak{M})$, and this completes the proof of the theorem.

5. *F. Bernstein's Equivalence-Theorem*

If two sets, $\mathfrak{M}$ and $\mathfrak{N}$, are given, precisely one of the following two cases can occur:

a) $\mathfrak{M}$ is equivalent to a subset, $\mathfrak{N}_1$, of $\mathfrak{N}$;

b) $\mathfrak{M}$ is equivalent to no subset of $\mathfrak{N}$.

Likewise, exactly one of the following two possibilities can take place:

α) $\mathfrak{N}$ is equivalent to a subset, $\mathfrak{M}_1$, of $\mathfrak{M}$;

β) $\mathfrak{N}$ is equivalent to no subset of $\mathfrak{M}$.

There are four conceivable combinations of these two pairs of cases, viz., aα, aβ, bα, bβ. In the two middle cases we have, according to p. 18, Def. 2, $|\mathfrak{M}| < |\mathfrak{N}|$, $|\mathfrak{M}| > |\mathfrak{N}|$, respectively. The occurrence of the last case, bβ, would mean that $\mathfrak{M}$ and $\mathfrak{N}$ are not comparable. As we have already remarked on p. 20, this case cannot take place, but the proof of this fact must be deferred until later. The case aα remains; and for it we have the following theorem, which was conjectured already by G. Cantor, and proved by F. Bernstein:

Equivalence Theorem. *If each of two sets, $\mathfrak{M}$ and $\mathfrak{N}$, is equivalent to a subset of the other, then $\mathfrak{M} \sim \mathfrak{N}$.*

This theorem can be reduced to the following proposition:

(P) If $\mathfrak{M}$ is equivalent to a subset $\mathfrak{M}_2$, then $\mathfrak{M}$ is also equivalent to every set $\mathfrak{M}_1$ "between" $\mathfrak{M}$ and $\mathfrak{M}_2$, i. e., to every set $\mathfrak{M}_1$ with the property that $\mathfrak{M}_2 \subseteq \mathfrak{M}_1 \subseteq \mathfrak{M}$.

In fact, suppose that $\mathfrak{M}_1$, $\mathfrak{N}_1$ are subsets of $\mathfrak{M}$, $\mathfrak{N}$ such that $\mathfrak{M} \sim \mathfrak{N}_1$, $\mathfrak{N} \sim \mathfrak{M}_1$. Then, by means of a mapping resulting from the last equivalence, the set $\mathfrak{N}$ is mapped on $\mathfrak{M}_1$, and hence, in particular, the subset $\mathfrak{N}_1$ is mapped on a subset, $\mathfrak{M}_2$, of $\mathfrak{M}_1$. Thus, $\mathfrak{M}_2 \subseteq \mathfrak{M}_1 \subseteq \mathfrak{M}$ and $\mathfrak{M} \sim \mathfrak{N}_1 \sim \mathfrak{M}_2$. Consequently, according to (P), $\mathfrak{M}_1 \sim \mathfrak{M}$; and since $\mathfrak{M}_1 \sim \mathfrak{N}$, we have also $\mathfrak{M} \sim \mathfrak{N}$.

For the proof of (P) we may obviously assume that $\mathfrak{M}_2 \subset \mathfrak{M}_1 \subset \mathfrak{M}$. $\mathfrak{M}$ is an infinite set, because $\mathfrak{M}_2 \sim \mathfrak{M}$. For convenience, set

$$\mathfrak{M}_2 = \mathfrak{A}, \quad \mathfrak{M}_1 - \mathfrak{M}_2 = \mathfrak{B}, \quad \mathfrak{M} - \mathfrak{M}_1 = \mathfrak{C}.$$

Then proposition (P) reads as follows:

(P*) If $\mathfrak{A}$, $\mathfrak{B}$, $\mathfrak{C}$ are disjunct sets, then

$$\mathfrak{A} + \mathfrak{B} + \mathfrak{C} \sim \mathfrak{A} \quad \text{implies} \quad \mathfrak{A} + \mathfrak{B} + \mathfrak{C} \sim \mathfrak{A} + \mathfrak{B}.$$

Now, according to the hypothesis that $\mathfrak{A} + \mathfrak{B} + \mathfrak{C} \sim \mathfrak{A}$, there exists a mapping φ of the set $\mathfrak{A} + \mathfrak{B} + \mathfrak{C}$ on $\mathfrak{A}$. Let $\mathfrak{A}_1$, $\mathfrak{B}_1$, $\mathfrak{C}_1$ be the subsets of $\mathfrak{A}$ which are the images of $\mathfrak{A}$, $\mathfrak{B}$,

$\mathfrak{C}$ (cf. the by all means very crude visualization through Fig. 5). Then

(1a) $$\mathfrak{A}_1 + \mathfrak{B}_1 + \mathfrak{C}_1 = \mathfrak{A},$$

(1b) $$\mathfrak{A} \sim \mathfrak{A}_1, \quad \mathfrak{B} \sim \mathfrak{B}_1, \quad \mathfrak{C} \sim \mathfrak{C}_1,$$

$\mathfrak{A}_1$, $\mathfrak{B}_1$, $\mathfrak{C}_1$ are disjunct.

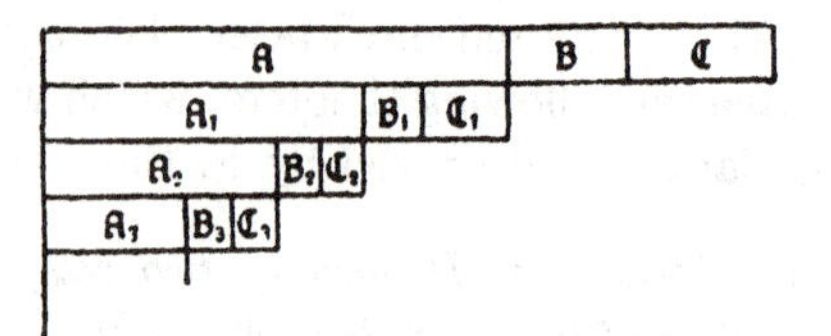

Fig. 5.

Since $\mathfrak{A}$ is mapped by φ on $\mathfrak{A}_1$, it follows from (1a) that the subsets $\mathfrak{A}_1$, $\mathfrak{B}_1$, $\mathfrak{C}_1$ of $\mathfrak{A}$ are mapped by φ on subsets $\mathfrak{A}_2$, $\mathfrak{B}_2$, $\mathfrak{C}_2$ of $\mathfrak{A}_1$. Then we have

(2a) $$\mathfrak{A}_2 + \mathfrak{B}_2 + \mathfrak{C}_2 = \mathfrak{A}_1,$$

(2b) $$\mathfrak{A}_1 \sim \mathfrak{A}_2, \quad \mathfrak{B}_1 \sim \mathfrak{B}_2, \quad \mathfrak{C}_1 \sim \mathfrak{C}_2,$$

$\mathfrak{A}_2$, $\mathfrak{B}_2$, $\mathfrak{C}_2$ are disjunct.

The next step leads to three sets $\mathfrak{A}_3$, $\mathfrak{B}_3$, $\mathfrak{C}_3$ with

(3a) $$\mathfrak{A}_3 + \mathfrak{B}_3 + \mathfrak{C}_3 = \mathfrak{A}_2,$$

(3b) $$\mathfrak{A}_2 \sim \mathfrak{A}_3, \quad \mathfrak{B}_2 \sim \mathfrak{B}_3, \quad \mathfrak{C}_2 \sim \mathfrak{C}_3,$$

$\mathfrak{A}_3$, $\mathfrak{B}_3$, $\mathfrak{C}_3$ are disjunct;

et cetera. Due to the fact that $\mathfrak{A} \sim \mathfrak{A}_1 \sim \mathfrak{A}_2 \sim \cdots$, the process does not terminate. Note especially the equivalence

(I) $$\mathfrak{C} \sim \mathfrak{C}_1 \sim \mathfrak{C}_2 \sim \cdots$$

arising from (1b), (2b), $\cdots$.

If we now set $\mathfrak{D} = \mathfrak{A}_1 \cdot \mathfrak{A}_2 \cdot \mathfrak{A}_3 \cdots$ ($\mathfrak{D}$ may be empty), then

$$\mathfrak{A} + \mathfrak{B} + \mathfrak{C} = \mathfrak{D} + \mathfrak{B} + \mathfrak{C} + \mathfrak{B}_1 + \mathfrak{C}_1 + \mathfrak{B}_2 + \mathfrak{C}_2 + \cdots,$$

$$\mathfrak{A} + \mathfrak{B} = \mathfrak{D} + \mathfrak{B} + \mathfrak{C}_1 + \mathfrak{B}_1 + \mathfrak{C}_2 + \mathfrak{B}_2 + \mathfrak{C}_3 + \cdots.$$

Here, on the right-hand side of the first equation, all the terms are mutually exclusive, and the same is true of the second equation. Hence, $\mathfrak{A} + \mathfrak{B}$ is mapped on $\mathfrak{A} + \mathfrak{B} + \mathfrak{C}$, and thereby (P*) is proved, if we succeed in establishing a mapping between every term on the right-hand side of the first equation, and the term directly below it in the second equation. The existence of this mapping, however, is ensured by (I).

An immediate consequence of the equivalence theorem is the following

Corollary. If $\mathfrak{M}$ *is equivalent to a subset of* $\mathfrak{N}$, *then* $|\mathfrak{M}| \leq |\mathfrak{N}|$.

6. *The Sum of Two Cardinal Numbers*

We must now develop further the operations on cardinal numbers. We shall first define addition.

A child, if asked to determine how many balls are three balls and two balls, will combine these two sets into a single one, and then ascertain the number of elements of this union of the two sets. With abstract finite sets, this method must be applied with caution. For, each of the sets $\{1, 2, 3\}$, $\{1, 2, 4\}$ has three elements. Their union, however, is $\{1, 2, 3, 4\}$. This set has only four elements, and so it cannot be used for determining the sum of the numbers 3 and 3. On the other hand, the method is always applicable to finite sets, if the sets representing the numbers are disjunct.

A suitable generalization of this method gives rise to the following general definition of the sum of two cardinal numbers. Let the cardinal numbers $\mathfrak{m}$ and $\mathfrak{n}$ be represented by the disjunct sets[3] $\mathfrak{M}$ and $\mathfrak{N}$. Form their union, $\mathfrak{S} = \mathfrak{M} + \mathfrak{N}$, and then stipulate that $\mathfrak{m} + \mathfrak{n} = |\mathfrak{S}|$.

For this definition to have a meaning, it is necessary to show that one arrives at the same cardinal number $|\mathfrak{S}|$ by starting

[3] Two cardinal numbers $\mathfrak{m}$ and $\mathfrak{n}$ can always be represented by disjunct sets. For, if $\mathfrak{M}(m)$ and $\mathfrak{N}(n)$ are sets having cardinal numbers $\mathfrak{m}$ and $\mathfrak{n}$, it is sufficient to replace every element m of the first set by m_1 or $(m, 1)$, and every element n of the second set by n_2 or $(n, 2)$. The sets obtained in this manner are certainly disjunct, and they too represent the cardinal numbers $\mathfrak{m}$ and $\mathfrak{n}$.

from two arbitrary sets $\overline{\mathfrak{M}}$, $\overline{\mathfrak{N}}$ which are equivalent to $\mathfrak{M}$, $\mathfrak{N}$, and which, of course, are likewise mutually exclusive. In fact, the mappings resulting from $\mathfrak{M} \sim \overline{\mathfrak{M}}$ and $\mathfrak{N} \sim \overline{\mathfrak{N}}$ then also furnish a mapping of the set $\mathfrak{M} + \mathfrak{N}$ on the set $\overline{\mathfrak{M}} + \overline{\mathfrak{N}}$.

Since the relations

$$\mathfrak{M} + \mathfrak{N} = \mathfrak{N} + \mathfrak{M},$$

$$(\mathfrak{M} + \mathfrak{N}) + \mathfrak{P} = \mathfrak{M} + (\mathfrak{N} + \mathfrak{P})$$

are valid for arbitrary sets, we have

α) $\mathfrak{m} + \mathfrak{n} = \mathfrak{n} + \mathfrak{m}$ (commitatuve law);

β) $(\mathfrak{m} + \mathfrak{n}) + \mathfrak{p} = \mathfrak{m} + (\mathfrak{n} + \mathfrak{p})$ (associative law).[4]

γ) Further, if

$$\mathfrak{m}_1 \leq \mathfrak{n}_1, \quad \mathfrak{m}_2 \leq \mathfrak{n}_2,$$

then invariably

$$\mathfrak{m}_1 + \mathfrak{m}_2 \leq \mathfrak{n}_1 + \mathfrak{n}_2 .$$

For if the four cardinal numbers are represented by the four disjunct sets $\mathfrak{M}_1$, $\mathfrak{M}_2$, $\mathfrak{N}_1$, $\mathfrak{N}_2$, then by hypothesis there exist subsets $\overline{\mathfrak{N}}_1 \subseteq \mathfrak{N}_1$ and $\overline{\mathfrak{N}}_2 \subseteq \mathfrak{N}_2$ such that $\mathfrak{M}_1 \sim \overline{\mathfrak{N}}_1$ and $\mathfrak{M}_2 \sim \overline{\mathfrak{N}}_2$. Since the sets are disjunct, the mappings resulting from these equivalences then also map the set $\mathfrak{M}_1 + \mathfrak{M}_2$ on $\overline{\mathfrak{N}}_1 + \overline{\mathfrak{N}}_2$. Consequently,

$$\mathfrak{M}_1 + \mathfrak{M}_2 \sim \overline{\mathfrak{N}}_1 + \overline{\mathfrak{N}}_2$$

and $\overline{\mathfrak{N}}_1 + \overline{\mathfrak{N}}_2 \subseteq \mathfrak{N}_1 + \mathfrak{N}_2$, and therefore, by the corollary of the equivalence theorem (p. 25), $\mathfrak{m}_1 + \mathfrak{m}_2 \leq \mathfrak{n}_1 + \mathfrak{n}_2$.

Remark 1. We cannot, however, always infer from $\mathfrak{m}_1 < \mathfrak{n}_1$ and $\mathfrak{m}_2 \leq \mathfrak{n}_2$, that $\mathfrak{m}_1 + \mathfrak{m}_2 < \mathfrak{n}_1 + \mathfrak{n}_2$, because, e.g., for every finite cardinal number n we have $n < \mathfrak{a}$ and $\mathfrak{a} \leq \mathfrak{a}$, but (cf. several lines below) $\mathfrak{a} + n = \mathfrak{a} + \mathfrak{a}$.

---

[4]On the basis of the associative law, a finite sum of more than two cardinal numbers can also be introduced. We shall postpone this generalization, as well as a much stronger one, until §8.

Remark 2. Whether the relations $\mathfrak{m}_1 < \mathfrak{n}_1$, $\mathfrak{m}_2 < \mathfrak{n}_2$ invariably imply

$$\mathfrak{m}_1 + \mathfrak{m}_2 < \mathfrak{n}_1 + \mathfrak{n}_2$$

cannot be decided until later (p. 123).

The following rules are valid for the addition of the simplest cardinal numbers:

a) $$\mathfrak{a} + n = \mathfrak{a}, \quad \mathfrak{a} + \mathfrak{a} = \mathfrak{a}.$$

For, the cardinal numbers n and $\mathfrak{a}$ can be represented by $\{1, 2, \cdots, n\}$, $\{n + 1, n + 2, \cdots\}$, respectively, and the sum of these sets is the set of natural numbers, which has power $\mathfrak{a}$.—To derive the second rule, represent the first term by the set of odd integers, and the second by the set of even integers. The sum of these sets is the enumerable set of all integers.

b) $$\mathfrak{c} + n = \mathfrak{c}, \quad \mathfrak{c} + \mathfrak{a} = \mathfrak{c}, \quad \mathfrak{c} + \mathfrak{c} = \mathfrak{c}.$$

To prove the last equation, represent the first term by the points of the interval $\langle 0, 1)$, the second by $\langle 1, 2\rangle$. Since $\langle 0, 1) + \langle 1, 2\rangle = \langle 0, 2\rangle$, which again has power $\mathfrak{c}$, the third equation is proved. From this one, the other two follow with the help of γ). For we have

$$\mathfrak{c} = \mathfrak{c} + 0 \leq \begin{Bmatrix} \mathfrak{c} + n \\ \mathfrak{c} + \mathfrak{a} \end{Bmatrix} \leq \mathfrak{c} + \mathfrak{c} = \mathfrak{c}.$$

Remark. Whether $\mathfrak{m} + \mathfrak{m} = \mathfrak{m}$ is true for every transfinite cardinal number $\mathfrak{m}$ cannot be settled until later (p. 122).

c) For every transfinite cardinal number $\mathfrak{m}$, $\mathfrak{m} + \mathfrak{a} = \mathfrak{m}$.

To demonstrate this, let $\mathfrak{m}$ and $\mathfrak{a}$ be represented by the disjunct sets $\mathfrak{M}$ and $\mathfrak{A}$. Since $\mathfrak{M}$ is an infinite set, it contains an enumerable subset $\mathfrak{B}$. $\mathfrak{A} + \mathfrak{B}$ is also enumerable, and can therefore be mapped on $\mathfrak{B}$; and, trivially, $\mathfrak{M} - \mathfrak{B}$ can be mapped on itself. $(\mathfrak{A} + \mathfrak{B}) + (\mathfrak{M} - \mathfrak{B})$ is thereby mapped on $\mathfrak{B} + (\mathfrak{M} - \mathfrak{B})$, i. e., $\mathfrak{M} + \mathfrak{A}$ is mapped on $\mathfrak{M}$, which proves the assertion.

Remark. It follows from c) that the set of transcendental

numbers has power $\mathfrak{c}$. For if $\mathfrak{M}$ is the set of transcendental numbers, and $\mathfrak{A}$ is the set of algebraic numbers, then $\mathfrak{M} + \mathfrak{A}$ is the set of all real numbers, and therefore has power $\mathfrak{c}$. Consequently, $|\mathfrak{M}| = |\mathfrak{M}| + |\mathfrak{A}| = |\mathfrak{M} + \mathfrak{A}| = \mathfrak{c}$.

7. *The Product of Two Cardinal Numbers*

For a child, multiplication is a repeated addition; e. g., $3 \cdot 4 = 4 + 4 + 4$. In order to represent each term, one has to write down four elements, say a, b, c, d, do this once more, then once again, and finally determine the total number of elements written down. Here the first a must be distinguished from the second and third. This distinction can be made by attaching an index, or by forming the sets $\{1, a\}$, $\{2, a\}$, $\{3, a\}$. We shall proceed with arbitrary sets in an analogous fashion, and we first make the following

Definition 1. *Let* $\mathfrak{M}$ *and* $\mathfrak{N}$ *be two disjunct sets with elements* m, n, *respectively. By the combinatorial product, or product of the second kind, of the two sets* $\mathfrak{M}$ *and* $\mathfrak{N}$, *in symbols:* $\mathfrak{M} \times \mathfrak{N}$, *we shall mean the set of all sets* $\{m, n\}$ *that can be formed from the elements* m *of* $\mathfrak{M}$ *and* n *of* $\mathfrak{N}$. *If at least one of the sets* $\mathfrak{M}$, $\mathfrak{N}$ *is empty, we stipulate that* $\mathfrak{M} \times \mathfrak{N} = 0$.

If, e. g., $\mathfrak{M} = \{1, 2, 3\}$ and $\mathfrak{N} = \{4, 5\}$, then

$$\mathfrak{M} \times \mathfrak{N} = \{\{1, 4\}, \{1, 5\}, \{2, 4\}, \{2, 5\}, \{3, 4\}, \{3, 5\}\}.$$

For finite sets $\mathfrak{M}$ and $\mathfrak{N}$, the cardinal number of $\mathfrak{M} \times \mathfrak{N}$ is obviously equal to the product of the cardinal numbers of $\mathfrak{M}$ and $\mathfrak{N}$. Hence we set down next

Definition 2. *Let the cardinal numbers* $\mathfrak{m}$ *and* $\mathfrak{n}$ *be represented by the disjunct sets* $\mathfrak{M}$ *and* $\mathfrak{N}$. *Then the product* $\mathfrak{m} \cdot \mathfrak{n}$ *shall equal* $|\mathfrak{M} \times \mathfrak{N}|$.

For disjunct sets $\mathfrak{M}$, $\mathfrak{N}$ and $\overline{\mathfrak{M}}$, $\overline{\mathfrak{N}}$, if $\mathfrak{M} \sim \overline{\mathfrak{M}}$ and $\mathfrak{N} \sim \overline{\mathfrak{N}}$, then $\mathfrak{M} \times \mathfrak{N} \sim \overline{\mathfrak{M}} \times \overline{\mathfrak{N}}$; i. e., the product $\mathfrak{m} \cdot \mathfrak{n}$ is independent of the particular representatives of the cardinal numbers $\mathfrak{m}$ and $\mathfrak{n}$. The proof of this is left to the reader.

In practice, the following form of the definition is often more convenient:

Definition 2a. *Let the cardinal numbers* $\mathfrak{m} > 0$ *and* $\mathfrak{n} > 0$ *be*

represented by the (not necessarily disjunct) sets $\mathfrak{M}(m)$ *and* $\mathfrak{N}(n)$. *Form the set* $\mathfrak{P}$ *of all element pairs* (m, n), *where the first place always contains an element of* $\mathfrak{M}$, *and the second, an element of* $\mathfrak{N}$. *Then* $\mathfrak{m} \cdot \mathfrak{n}$ *shall equal* $|\mathfrak{P}|$. *If at least one of the cardinal numbers* $\mathfrak{m}$, $\mathfrak{n}$ *is zero, we set* $\mathfrak{m} \cdot \mathfrak{n} = 0$.

That this definition gives the same result as the previous one follows from the fact that disjunct sets are obtained from $\mathfrak{M}(m)$ and $\mathfrak{N}(n)$ by attaching the index 1 to every m and the index 2 to every n, and that the set of all element pairs (m, n) is equivalent to the set of all sets $\{m_1, n_2\}$.

It is obvious that $\mathfrak{M} \times \mathfrak{N} = \mathfrak{N} \times \mathfrak{M}$, and hence

$$\alpha) \qquad \mathfrak{m} \cdot \mathfrak{n} = \mathfrak{n} \cdot \mathfrak{m} \qquad \textit{(commutative law)}.$$

We have, further, $(\mathfrak{M} \times \mathfrak{N}) \times \mathfrak{P} \sim \mathfrak{M} \times (\mathfrak{N} \times \mathfrak{P})$. To see this, for every triple of elements m, n, p of these sets simply let the element $\{\{m, n\}, p\}$ of the first combinatorial product correspond to the element $\{m, \{n, p\}\}$ of the second. Consequently,

$$\beta) \qquad (\mathfrak{m} \cdot \mathfrak{n}) \cdot \mathfrak{p} = \mathfrak{m} \cdot (\mathfrak{n} \cdot \mathfrak{p}) \qquad \textit{(associative law)}.$$

For disjunct sets $\mathfrak{M}$, $\mathfrak{N}$, $\mathfrak{P}$, it is clear that $(\mathfrak{M} + \mathfrak{N}) \times \mathfrak{P} = \mathfrak{M} \times \mathfrak{P} + \mathfrak{N} \times \mathfrak{P}$. Hence,

$$\gamma) \qquad (\mathfrak{m} + \mathfrak{n})\mathfrak{p} = \mathfrak{m}\mathfrak{p} + \mathfrak{n}\mathfrak{p} \quad \textit{(distributive law)}.$$

δ) From
$$\mathfrak{m}_1 \leq \mathfrak{n}_1, \quad \mathfrak{m}_2 \leq \mathfrak{n}_2$$
it follows that
$$\mathfrak{m}_1 \cdot \mathfrak{m}_2 \leq \mathfrak{n}_1 \cdot \mathfrak{n}_2 .$$

For if the four cardinal numbers are represented by the four mutually exclusive sets $\mathfrak{M}_1$, $\mathfrak{M}_2$, $\mathfrak{N}_1$, $\mathfrak{N}_2$, there exist subsets $\overline{\mathfrak{N}}_1 \subseteq \mathfrak{N}_1$ and $\overline{\mathfrak{N}}_2 \subseteq \mathfrak{N}_2$ such that $\mathfrak{M}_1 \sim \overline{\mathfrak{N}}_1$ and $\mathfrak{M}_2 \sim \overline{\mathfrak{N}}_2$. But then obviously

$$\mathfrak{M}_1 \times \mathfrak{M}_2 \sim \overline{\mathfrak{N}}_1 \times \overline{\mathfrak{N}}_2$$

and $\overline{\mathfrak{N}}_1 \times \overline{\mathfrak{N}}_2 \subseteq \mathfrak{N}_1 \times \mathfrak{N}_2$, so that

$$|\mathfrak{M}_1 \times \mathfrak{M}_2| \leq |\mathfrak{N}_1 \times \mathfrak{N}_2|.$$

Remark. Here again we cannot infer from $\mathfrak{m}_1 < \mathfrak{n}_1$, $\mathfrak{m}_2 \leq \mathfrak{n}_2$, that $\mathfrak{m}_1\mathfrak{m}_2 < \mathfrak{n}_1\mathfrak{n}_2$. Whether this inequality is implied by $\mathfrak{m}_1 < \mathfrak{n}_1$, $\mathfrak{m}_2 < \mathfrak{n}_2$, cannot be decided until p. 123.

We shall now become acquainted with the beginning of the multiplication table for cardinal numbers.

a) $n \cdot \mathfrak{a} = \mathfrak{a}$ for every finite number $n \neq 0$; $\mathfrak{a} \cdot \mathfrak{a} = \mathfrak{a}$.

The second rule is derived as follows: Let $\mathfrak{a}$ be represented by the set $\{1, 2, 3, \cdots\}$. Then, according to Definition 2a, the product $\mathfrak{a} \cdot \mathfrak{a}$ is represented by the set of all number pairs (a, b), i. e., by the set

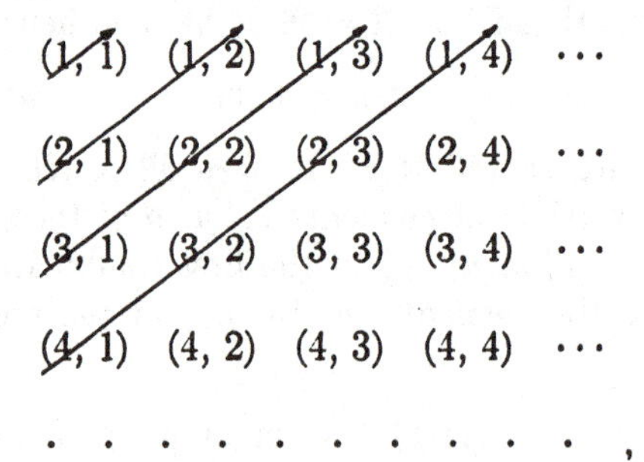

which can again be enumerated in the indicated manner by the first diagonal method. From this it follows, further, using δ), that $\mathfrak{a} \leq n \cdot \mathfrak{a} \leq \mathfrak{a} \cdot \mathfrak{a} = \mathfrak{a}$; i. e., the first rule is also valid.

b) $n \cdot \mathfrak{c} = \mathfrak{c}$ for every finite number $n \neq 0$; $\mathfrak{a} \cdot \mathfrak{c} = \mathfrak{c}$.

To prove the second rule, represent $\mathfrak{a}$ by the set $\{1, 2, 3, \cdots\}$, and $\mathfrak{c}$ by the point set $\langle 0, 1)$. Then the combinatorial product is the set of all sets $\{n, x\}$, where n is an arbitrary natural number, and x is an arbitrary point of the interval $\langle 0, 1)$. If each of the sets $\{n, x\}$ is made to correspond to the point $n + x$, we obtain a mapping of the combinatorial product on the half-line which begins at 1, and this half-line is a point set having power $\mathfrak{c}$. This proves the second rule. The first follows from the fact that $\mathfrak{c} \leq n \cdot \mathfrak{c} \leq \mathfrak{a} \cdot \mathfrak{c} = \mathfrak{c}$.

c) $$\mathfrak{c} \cdot \mathfrak{c} = \mathfrak{c}.$$

For the proof, let $\mathfrak{c}$ be represented by the numbers of the interval $(0, 1\rangle$. Then $\mathfrak{c} \cdot \mathfrak{c}$ is represented by the set of number pairs (x, y) with $0 < x \leq 1, 0 < y \leq 1$. The assertion is proved as soon as we show that the set of these number pairs can be mapped on the interval $0 < z \leq 1$. For this demonstration, represent each of the three numbers x, y, z by a nonterminating decimal fraction. Since the representation of our numbers by means of such decimal fractions is a unique one, it is necessary to show merely that the set of decimal-fraction pairs x, y can be mapped on the set of decimal fractions z. Let the pair of decimal fractions x, y be given; e. g.,

$$x = 0.3 \quad 06 \quad 007 \quad 8 \quad 9 \quad \cdots,$$

$$y = 0.001 \quad 2 \quad 3 \quad 004 \quad 6 \quad \cdots.$$

In these we mark off numeral complexes, by going always up to the next nonzero numeral, inclusive. Now we write down the first x-complex, after that the first y-complex, then the second x-complex, followed by the second y-complex, etc. Since neither x nor y exhibits only zeros from a certain point on, the process can be continued without ending, and gives rise to a nonterminating decimal-fraction; in our example, to

$$z = 0.3 \quad 001 \quad 06 \quad 2 \quad 007 \quad 3 \quad 8 \quad 004 \quad 9 \quad 6 \quad \cdots.$$

The decimal fraction arising in this manner shall be made to correspond to the number pair x, y. Every number pair x, y thus determines precisely one z, and we see at the same time, that the complexes of x, y also yield just the complexes of z. From this it follows that, conversely, every z determines exactly one number pair x, y, and, in fact, precisely that pair which gave rise to z; we have only to mark off the complexes in z, and arrange them alternately so as to form two decimal fractions. A one-to-one correspondence between the number pairs x, y and the numbers z is thereby established, and this completes the proof of the rule.

Rule c) contains a fact which at first appears downright paradoxical. If we interpret the number pairs x, y as rectangular coordinates of points of the plane, then it follows from c) that the points of a square can be mapped on the segment $\langle 0, 1 \rangle$.

What is more, we can represent the factors $\mathfrak{c}$ by the set of *all* real numbers. Then the number pairs x, y yield, in fact, the points of the whole plane, and we can now infer from rule c) that the entire plane can be mapped on the segment $\langle 0, 1\rangle$. This means that there exist functions $f(z)$ and $g(z)$ such that the number pairs $x = f(z)$, $y = g(z)$ yield the coordinates of all the points of the plane as z ranges over the interval $\langle 0, 1\rangle$, and that every point of the plane is thereby obtained only once. Thus, the dimension number of a geometrical configuration need not be preserved under such mappings. The dimension number is, however, invariant under one-to-one mappings, if these are, in addition, continuous. For a general proof of this theorem, cf. L. E. J. Brouwer, Math. Ann. 70 (1911), 161–165.

8. *The Sum of Arbitrarily Many Cardinal Numbers*

Let there be given a nonempty set $\mathfrak{K}(k)$, and to every element k of this set let there correspond a set $\mathfrak{M}_k$. We shall put this more briefly by saying: Let there be given a set complex $\mathfrak{K}(\mathfrak{M}_k)$ arising from $\mathfrak{K}(k)$.[5] A cardinal-number complex $\mathfrak{K}(\mathfrak{m}_k)$ is defined analogously. The union of all the sets $\mathfrak{M}_k$ of the set complex $\mathfrak{K}(\mathfrak{M}_k)$ shall be denoted by $\sum_{k\in\mathfrak{K}} \mathfrak{M}_k$. This is a generalization of the notation introduced on p. 7 only for enumerably many sets.

The sum of arbitrarily many cardinal numbers can now be defined as follows:

Let $\mathfrak{K}(\mathfrak{m}_k)$ be a cardinal-number complex arising from the set $\mathfrak{K}(k)$. Let every cardinal number $\mathfrak{m}_k$ be represented by a set $\mathfrak{M}_k$ in such a manner, that all the $\mathfrak{M}_k$'s are mutually exclusive.[6] Then the sum of the cardinal-number complex shall be defined as

$$\sum_{k\in\mathfrak{K}} \mathfrak{m}_k = \left| \sum_{k\in\mathfrak{K}} \mathfrak{M}_k \right|.$$

---

[5]Since the same set may correspond to distinct elements k, a set *complex* need not be a *set* of sets.

[6]It is sufficient, in order to obtain such a representation, to furnish the elements m of the set $\mathfrak{M}_k$ with the index k.

It again follows immediately from the representation of the cardinal numbers by disjunct sets $\mathfrak{M}_k$, that the sum of the cardinal numbers is independent of their particular representatives. Further, the above definition obviously contains as a special case the one in §6.

The three rules of operation which hold for sums of two sets can also be generalized. With regard to the commutative law, it is to be noted that the fact that the sum is independent of the order of succession of the terms is already included in the definition, since no order of succession of the elements of the set $\mathfrak{K}(k)$ is specified. The commutative law therefore assumes a form which differs somewhat from the usual one.

THEOREM 1 (commutative law). *Two cardinal-number complexes $\mathfrak{K}(\mathfrak{m}_k)$ and $\mathfrak{L}(\mathfrak{n}_l)$ have the same sum, if every cardinal number $\mathfrak{p}$ appearing in the one complex also appears, and with the same multiplicity, in the other complex. Precisely formulated, this means the following: Let $\mathfrak{K}_\mathfrak{p}(k)$ denote the subset of elements k of $\mathfrak{K}(k)$ to which the same cardinal number $\mathfrak{p}$ corresponds. Then the different $\mathfrak{K}_\mathfrak{p}(k)$'s are mutually exclusive. Define the sets $\mathfrak{L}_\mathfrak{p}(l)$ in like manner. Then*

$$\sum_{\mathfrak{p}} \mathfrak{K}_\mathfrak{p}(k) = \mathfrak{K}(k), \qquad \sum_{\mathfrak{p}} \mathfrak{L}_\mathfrak{p}(l) = \mathfrak{L}(l).$$

It is assumed, now, that, for every $\mathfrak{p}$,

$$\mathfrak{K}_\mathfrak{p}(k) \sim \mathfrak{L}_\mathfrak{p}(l), \tag{1}$$

and it is asserted that

$$\sum_{k \in \mathfrak{K}} \mathfrak{m}_k = \sum_{l \in \mathfrak{L}} \mathfrak{n}_l .$$

Proof: Represent all the cardinal numbers $\mathfrak{m}_k$, corresponding to the elements k of $\mathfrak{K}(k)$, by disjunct sets $\mathfrak{M}_k$; and all the cardinal numbers $\mathfrak{n}_l$, corresponding to the elements l of $\mathfrak{L}(l)$, by disjunct sets $\mathfrak{N}_l$. Because of (1), there is a one-to-one correspondence between the elements k and l of each pair of sets $\mathfrak{K}_\mathfrak{p}(k)$ and $\mathfrak{L}_\mathfrak{p}(l)$. Keep this mapping fixed. Suppose an element l corresponds to the element k. Then to these elements correspond sets $\mathfrak{M}_k$ and $\mathfrak{N}_l$ having the same power $\mathfrak{p}$. These sets therefore can also be mapped on one another. This, however, determines a one-to-one mapping of the set $\sum_{k \in \mathfrak{K}} \mathfrak{M}_k$ on the set $\sum_{l \in \mathfrak{L}} \mathfrak{N}_l$. For if m is an arbitrary element of the first of these two sets, it belongs to a well-determined $\mathfrak{M}_k$, and consequently a well-determined n of the second sum corresponds to it in a fully unique manner; and conversely.

In connection with the general formulation of the associative law, the following is to be noted. In its simplest form, this law asserts that $(\mathfrak{m} + \mathfrak{n}) + \mathfrak{p} = \mathfrak{m} + (\mathfrak{n} + \mathfrak{p})$, i. e., that, for the addition of three cardinal numbers, it is immaterial how this set of three elements is decomposed into subsets of at most two elements, in order to calculate the sums successively. Since we now have a general definition of sum for arbitrarily many terms, the associative law can immediately be given the form

$$(\mathfrak{m} + \mathfrak{n}) + \mathfrak{p} = \mathfrak{m} + \mathfrak{n} + \mathfrak{p}, \quad \mathfrak{n} + (\mathfrak{m} + \mathfrak{p}) = \mathfrak{m} + \mathfrak{n} + \mathfrak{p},$$

etc. Thus we arrive at

THEOREM 2 (associative law). *To every element l of a set $\mathfrak{L}(l)$, let there correspond a set $\mathfrak{K}_l(k)$. These sets $\mathfrak{K}_l(k)$ are assumed to be mutually exclusive. Let $\mathfrak{K}(k) = \sum_{l \in \mathfrak{L}} \mathfrak{K}_l(k)$, and to every k of $\mathfrak{K}(k)$ let there correspond a cardinal number $\mathfrak{m}_k$. Then a cardinal-number complex $\mathfrak{K}(\mathfrak{m}_k)$ and a set of cardinal-number complexes $\mathfrak{K}_l(\mathfrak{m}_k)$ are defined. The assertion, now, is that*

$$\sum_{l \in \mathfrak{L}} \sum_{k \in \mathfrak{K}_l} \mathfrak{m}_k = \sum_{k \in \mathfrak{K}} \mathfrak{m}_k .$$

Proof: Once again we represent the cardinal numbers $\mathfrak{m}_k$ by the disjunct sets $\mathfrak{M}_k$. Then $\sum_{k \in \mathfrak{K}_l} \mathfrak{M}_k$ represents the cardinal number $\sum_{k \in \mathfrak{K}_l} \mathfrak{m}_k$, and the assertion now follows immediately from the obviously correct equation

$$\sum_{l \in \mathfrak{L}} \sum_{k \in \mathfrak{K}_l} \mathfrak{M}_k = \sum_{k \in \mathfrak{K}} \mathfrak{M}_k .$$

THEOREM 3. *Let $\mathfrak{K}(\mathfrak{m}_k)$ and $\mathfrak{K}(\mathfrak{n}_k)$ be two cardinal-number complexes arising from the set $\mathfrak{K}(k)$, and let $\mathfrak{m}_k \leq \mathfrak{n}_k$ for every k. Then*

$$\sum_{k \in \mathfrak{K}} \mathfrak{m}_k \leq \sum_{k \in \mathfrak{K}} \mathfrak{n}_k .$$

Proof: Represent the cardinal numbers $\mathfrak{m}_k$ by disjunct sets $\mathfrak{M}_k$, and the cardinal numbers $\mathfrak{n}_k$ by disjunct sets $\mathfrak{N}_k$. We have to show that $\sum \mathfrak{M}_k$ is equivalent to a subset of $\sum \mathfrak{N}_k$. This follows immediately, however, from the fact that, by hypothesis, every $\mathfrak{M}_k$ is equivalent to a subset of $\mathfrak{N}_k$. Keep fixed a mapping of the set $\mathfrak{M}_k$ on a part of $\mathfrak{N}_k$, resulting from this equivalence. Then, because of the mutual exclusiveness of the sets $\mathfrak{M}_k$, and that of the sets $\mathfrak{N}_k$, this gives at once a mapping of the set $\sum \mathfrak{M}_k$ on a subset of $\sum \mathfrak{N}_k$.

Let us prove another inequality concerning cardinal numbers.

THEOREM 4. *Let $\mathfrak{K}(\mathfrak{m}_k)$ be a cardinal-number complex which contains no greatest cardinal number. Then, for every cardinal number $\mathfrak{m}$ of the complex,*

$$\sum_{k \in \mathfrak{K}} \mathfrak{m}_k > \mathfrak{m}.$$

Proof: If we replace by 0 all the terms in the sum of the complex except one, $\mathfrak{m}_l$, then Theorem 3 states that

$$\sum_{k \in \mathfrak{K}} \mathfrak{m}_k \geq \mathfrak{m}_l .$$

If, for some l, the equality sign held here, we should have, for this l_0, that every $\mathfrak{m}_l \leq \mathfrak{m}_{l_0}$. Thus there would be a greatest among the cardinal numbers of the complex, namely, $\mathfrak{m}_{l_0}$, contrary to hypothesis.

Remark. Note the following: It is not assumed in Theorem 4, that to every cardinal number of the complex there is a still greater cardinal number within the complex. The hypothesis of the theorem would still be fulfilled if, e. g., the complex contained a cardinal number which was not comparable with any other number of the complex.

THEOREM 5. *For every set $\mathfrak{M}$ of cardinal numbers there exists a cardinal number which is greater than every cardinal number in $\mathfrak{M}$.*

Proof: If there is a greatest cardinal number among the elements of $\mathfrak{M}$, then, by the theorem on p. 21, there exists a cardinal number greater than this one. If there is no greatest cardinal number among the elements of $\mathfrak{M}$, the assertion follows from Theorem 4.

From Theorem 5 follows a fact which is of very great importance for the theory of sets. According to this theorem, namely, there is no set which contains *all* cardinal numbers; the "set of all cardinal numbers" is therefore a meaningless concept. This comes as a great surprise after the procedure we have been following thus far, because up to now it has been regarded as self-evident that the formation of sets is subject to no restrictions. The result just obtained shows that such an unrestrained formation of sets may lead to contradictions. A discussion of this first antinomy, as well as others, will be postponed until we have a more extensive fund of the results and arguments of the theory of sets. Cf. the concluding remarks on p. 135.

The product of two cardinal numbers $\mathfrak{k} \cdot \mathfrak{m}$ can now be regarded also as a "sum of $\mathfrak{k}$ cardinal numbers $\mathfrak{m}$". For if the set $\mathfrak{K}(k)$ has power $\mathfrak{p}$, and if to every element of $\mathfrak{K}(k)$ we make correspond the same cardinal number $\mathfrak{m}$, the sum of this complex is precisely a sum of $\mathfrak{k}$ equal terms $\mathfrak{m}$. According to the definition of sum, the $\mathfrak{m}$'s corresponding to the different elements k must be represented by mutually exclusive sets. This can be effected by taking an arbitrary set $\mathfrak{M}(m)$ with power $\mathfrak{m}$ and with no element in common with $\mathfrak{K}$, and letting correspond to every element k the set $\mathfrak{M}(\{k, m\})$. Then the set $\sum_{k \in \mathfrak{K}} \mathfrak{M}(\{k, m\})$ represents, on the one hand, the sum of $\mathfrak{k}$ terms $\mathfrak{m}$, and, on the other hand, is equal to $\mathfrak{K} \times \mathfrak{M}$, so that it also represents the cardinal number $\mathfrak{k} \cdot \mathfrak{m}$.

The proof of the equation $\mathfrak{a} \cdot \mathfrak{c} = \mathfrak{c}$ on p. 30 really rests on this connection between addition and multiplication. Further, it now follows immediately from the results of §7, that a sum of enumerably many ones gives the cardinal number $\mathfrak{a}$, and a sum of $\mathfrak{c}$ ones yields the cardinal number $\mathfrak{c}$. Finally, a sum of enumerably many cardinal numbers $\mathfrak{a}$ is also equal to $\mathfrak{a} \cdot \mathfrak{a} = \mathfrak{a}$. From these results we obtain, e. g.: $1 + 2 + 3 + \cdots = \mathfrak{a}$; for we have

$$\mathfrak{a} = 1 + 1 + 1 + \cdots \leq 1 + 2 + 3 + \cdots$$

$$\leq \mathfrak{a} + \mathfrak{a} + \mathfrak{a} + \cdots = \mathfrak{a} \cdot \mathfrak{a} = \mathfrak{a}.$$

Moreover,

$$1 + \mathfrak{a} + \mathfrak{c} + 2 + \mathfrak{a} + \mathfrak{c} + 3 + \mathfrak{a} + \mathfrak{c} + \cdots$$

$$= (1 + 2 + 3 + \cdots) + (\mathfrak{a} + \mathfrak{a} + \mathfrak{a} + \cdots)$$

$$+ (\mathfrak{c} + \mathfrak{c} + \mathfrak{c} + \cdots)$$

$$= \mathfrak{a} + \mathfrak{a} \cdot \mathfrak{a} + \mathfrak{a} \cdot \mathfrak{c} = \mathfrak{a} + \mathfrak{a} + \mathfrak{c} = \mathfrak{c}.$$

9. The Product of Arbitrarily Many Cardinal Numbers

Since the definition of product will again be based on the combinatorial product, we first generalize the latter notion.

Let a set $\mathfrak{K}(k)$ be given, and to every one of its elements k let there correspond a set $\mathfrak{M}_k$. These sets $\mathfrak{M}_k$ are assumed to be disjunct. From the set of sets $\mathfrak{K}(\mathfrak{M}_k)$, form a new set by replacing every set $\mathfrak{M}_k$ by an arbitrary one of its elements m_k. The sets $\mathfrak{K}(m_k)$ arising in this manner shall be taken as elements of a new set, called the combinatorial product, or product of the second kind, of the sets $\mathfrak{M}_k$; in symbols: ${}^{\times}\prod_{k \in \mathfrak{K}} \mathfrak{M}_k$. If at least one of the sets $\mathfrak{M}_k$ is empty, the product shall be 0.

If, e. g., $\mathfrak{K}$ is enumerable, and $\mathfrak{M}_1 = \{a, b\}$, $\mathfrak{M}_2 = \{c, d\}$, $\mathfrak{M}_3 = \{3\}$, $M_4 = \{4\}$, $\cdots$, then the combinatorial product is

$$\{\{a, c, 3, 4, \cdots\}, \{a, d, 3, 4, \cdots\},$$

$$\{b, c, 3, 4, \cdots\}, \{b, d, 3, 4, \cdots\}\}.$$

Note that the elements of the union of given sets are invariably *elements* of these sets, whereas the elements of the combinatorial product are always *sets of elements* of the given sets.

Now let $\mathfrak{K}(\mathfrak{m}_k)$ be a given cardinal-number complex arising from $\mathfrak{K}(k)$; i. e., to every element k of $\mathfrak{K}(k)$ let there correspond a cardinal number $\mathfrak{m}_k$. Represent each of the cardinal numbers $\mathfrak{m}_k$ by a set $\mathfrak{M}_k$, and in such a manner, that the $\mathfrak{M}_k$'s are mutually exclusive. Then the product of the cardinal-number complex shall be defined as

$$\prod_{k \in \mathfrak{K}} \mathfrak{m}_k = | {}^{\times}\prod_{k \in \mathfrak{K}} \mathfrak{M}_k |.$$

It is clear that this definition includes as a special case the one given in §7 for two factors. We must now show that the product of the cardinal numbers is independent of their particular representatives. To this end, let the $\overline{\mathfrak{M}}_k$'s be sets which are equivalent to the $\mathfrak{M}_k$'s, and keep fixed a one-to-one correspondence between their elements m and $\overline{m}$ which arises from this equivalence. In each $\mathfrak{K}(m_k)$, which is an element of the combinatorial product of the $\mathfrak{M}_k$'s, replace every element m by the element $\overline{m}$ corresponding to it. The result is an element $\mathfrak{K}(\overline{m}_k)$ of the combinatorial product of the $\overline{\mathfrak{M}}_k$'s, and if we let it correspond to $\mathfrak{K}(m_k)$, we have the desired equivalence between the two combinatorial products.

Theorem 1 (commutative law). *Two cardinal-number complexes $\mathfrak{K}(\mathfrak{m})$ and $\mathfrak{L}(\mathfrak{n})$ have the same product, if every cardinal number appearing in the one complex also appears, and with the same multiplicity, in the other complex. Precisely formulated, this means the following: Let $\mathfrak{K}_{\mathfrak{p}}(k)$ denote the subset of elements k of $\mathfrak{K}(k)$ to which the same cardinal number $\mathfrak{p}$ corresponds. Then the different $\mathfrak{K}_{\mathfrak{p}}(k)$'s are mutually exclusive. Define the sets $\mathfrak{L}_{\mathfrak{p}}(l)$ in like manner. Then*

$$\sum_{\mathfrak{p}} \mathfrak{K}_{\mathfrak{p}}(k) = \mathfrak{K}(k), \qquad \sum_{\mathfrak{p}} \mathfrak{L}_{\mathfrak{p}}(l) = \mathfrak{L}(l).$$

It is assumed, now, that, for every $\mathfrak{p}$,

$$\mathfrak{K}_{\mathfrak{p}}(k) \sim \mathfrak{L}_{\mathfrak{p}}(l), \tag{1}$$

and it is asserted that

$$\prod_{k \in \mathfrak{K}} \mathfrak{m}_k = \prod_{l \in \mathfrak{L}} \mathfrak{n}_l \,.$$

Proof: Represent all the cardinal numbers $\mathfrak{m}_k$, corresponding to the elements k of $\mathfrak{K}(k)$, by disjunct sets $\mathfrak{M}_k$; and all the cardinal numbers $\mathfrak{n}_l$, corresponding to the elements l of $\mathfrak{L}(l)$, by disjunct sets $\mathfrak{N}_l$. Because of (1), there is a one-to-one correspondence between the elements k and l of each pair of sets $\mathfrak{K}_\nu(k)$ and $\mathfrak{L}_\nu(l)$. Keep this mapping fixed. To the element k, let there correspond an element $l = \varphi(k)$. To these elements then correspond sets $\mathfrak{M}_k$ and $\mathfrak{N}_l$ of the same power. These sets can therefore be mapped on one another. Under this mapping, let the element $n_l = \psi(m_k)$ correspond to the element m_k. Keep this mapping fixed too. Now, in every element $\mathfrak{K}(m_k)$ of the combinatorial product of the $\mathfrak{M}_k$'s, replace each m_k by $n_l = \psi(m_k)$. Since every k belongs to exactly one $\mathfrak{K}_\nu(k)$, and every l belongs to exactly one $\mathfrak{L}_\nu(l)$, we obtain in this manner precisely one element $\mathfrak{L}(n_l)$ of the combinatorial product of the $\mathfrak{N}_l$'s. These elements $\mathfrak{K}(m_k)$ and $\mathfrak{L}(n_l)$ are now made to correspond to each other. This determines a mapping between the combinatorial products of the $\mathfrak{M}_k$'s and the $\mathfrak{N}_l$'s, and hence the assertion is proved.

THEOREM 2 (associative law). *To every element l of a set $\mathfrak{L}(l)$, let there correspond a set $\mathfrak{K}_l(k)$. These sets $\mathfrak{K}_l(k)$ are assumed to be mutually exclusive. Let $\mathfrak{K}(k) = \sum_{l \in \mathfrak{L}} \mathfrak{K}_l(k)$, and to every k of $\mathfrak{K}(k)$ let there correspond a cardinal number $\mathfrak{m}_k$. Then a cardinal-number complex $\mathfrak{K}(\mathfrak{m}_k)$ and a set of cardinal-number complexes $\mathfrak{K}_l(\mathfrak{m}_k)$ are defined. The assertion, now, is that*

$$\prod_{l \in \mathfrak{L}} \prod_{k \in \mathfrak{K}_l} \mathfrak{m}_k = \prod_{k \in \mathfrak{K}} \mathfrak{m}_k . \tag{2}$$

Proof: Represent the cardinal numbers $\mathfrak{m}_k$ by disjunct sets $\mathfrak{M}_k$. Then the product $\prod_{k \in \mathfrak{K}_l} \mathfrak{m}_k$ is represented by the set with the elements $\mathfrak{K}_l(\mathfrak{m}_k)$. For every $l \in \mathfrak{L}$ we get a combinatorial product with these elements, and the sets which we obtain for the different l's are mutually exclusive. Hence, the left-hand side of (2) is represented by a set whose elements are $\mathfrak{L}(\mathfrak{K}_l(m_k))$, i. e., by a set whose elements result from $\mathfrak{L}(l)$ by replacing every l by a set $\mathfrak{K}_l(m_k)$. Each $\mathfrak{L}(\mathfrak{K}_l(m_k))$ contains exactly one element of every $\mathfrak{M}_k$, and therefore determines precisely one element $\mathfrak{K}(m_k)$ of the combinatorial product of all the $\mathfrak{M}_k$'s. Let this element $\mathfrak{K}(m_k)$ correspond, now, to $\mathfrak{L}(\mathfrak{K}_l(m_k))$. The set which represents the left-hand side of (2) is thereby mapped on the combinatorial product of all the $\mathfrak{M}_k$'s, and the validity of (2) is thus established.

The general form of the distributive law has hardly any application, and will therefore not be formulated here.

THEOREM 3. *Let* $\mathfrak{K}(\mathfrak{m}_k)$ *and* $\mathfrak{K}(\mathfrak{n}_k)$ *be two cardinal-number complexes arising from the set* $\mathfrak{K}(k)$, *and let* $\mathfrak{m}_k \leq \mathfrak{n}_k$ *for every* k. *Then*

$$\prod_{k \in \mathfrak{K}} \mathfrak{m}_k \leq \prod_{k \in \mathfrak{K}} \mathfrak{n}_k .$$

Proof: Represent the cardinal numbers $\mathfrak{m}_k$ by disjunct sets $\mathfrak{M}_k$, and the $\mathfrak{n}_k$'s by disjunct sets $\mathfrak{N}_k$. We are to show that the combinatorial product of the $\mathfrak{M}_k$'s is equivalent to a subset of the combinatorial product of the $\mathfrak{N}_k$'s. By hypothesis, every $\mathfrak{M}_k$ is equivalent to a subset $\overline{\mathfrak{N}}_k$ of $\mathfrak{N}_k$. This gives rise to a mapping of $\mathfrak{M}_k$ on $\overline{\mathfrak{N}}_k$. Keep this mapping fixed. To each element $\mathfrak{K}(m_k)$ of the combinatorial product of all the $\mathfrak{M}_k$'s we now let correspond that $\mathfrak{K}(\overline{n}_k)$ which results from $\mathfrak{K}(m_k)$ by replacing every m_k by the $\overline{n}_k$ corresponding to it. The combinatorial product of the $\mathfrak{M}_k$'s is thereby mapped on the combinatorial product of the $\overline{\mathfrak{N}}_k$'s, i. e., on a subset of the combinatorial product of the $\mathfrak{N}_k$'s, which proves the assertion.

The principal application of the general definition of product is to introduce the power. We conclude this paragraph by listing merely a few simple consequences of, and remarks concerning, what has preceded; their justification may be left to the reader.

a) A sum of cardinal numbers is equal to zero, if, and only if, each of the terms is zero.

b) For every cardinal number $\mathfrak{m}$ we have $\mathfrak{m} + 0 = \mathfrak{m}$, and 0 is the only number which possesses this property for *every* number $\mathfrak{m}$.

c) A product of cardinal numbers is equal to zero, if, and only if, at least one of the factors is zero.

d) A product of cardinal numbers is equal to 1, if, and only if, each of the factors is equal to 1.

e) For every cardinal number $\mathfrak{m}$ we have $\mathfrak{m} \cdot 1 = \mathfrak{m}$, and 1 is the only number which possesses this property for *every* number $\mathfrak{m}$.

f) One might try to define subtraction and division of cardinal numbers, in order to proceed from these perhaps even to an introduction of negative and fractional cardinal numbers. For natural numbers, the difference $n - m$, or the quotient n/m, is defined as the solution, if it exists, of the equation $m + x = n$, or $m \cdot x = n$, respectively, (the latter only for $m \neq 0$). This definition is possible here because each of these equations has at most one solution in the domain of natural numbers. If one attempts to imitate this in the domain of cardinal numbers, it fails, because these equations can have actually infinitely many solutions in the domain of cardinal numbers, as the following examples show:

$$\mathfrak{a} + 1 = \mathfrak{a} + 2 = \mathfrak{a} + 3 = \cdots = \mathfrak{a};$$

$$\mathfrak{a} \cdot 1 = \mathfrak{a} \cdot 2 = \mathfrak{a} \cdot 3 = \cdots = \mathfrak{a}.$$

10. The Power

The introduction of the power, which leads to some quite interesting results, will be carried out first according to the pattern of ordinary arithmetic. Let $\mathfrak{k} \neq 0$ and $\mathfrak{m}$ be two arbitrary cardinal numbers. Then $\mathfrak{m}^{\mathfrak{k}}$ is defined as a product of $\mathfrak{k}$ factors $\mathfrak{m}$. When formulated in greater detail, this means the following: Let there be given a set $\mathfrak{K}(k)$ having cardinal number $\mathfrak{k}$. With every element k we associate the same cardinal number $\mathfrak{m}$. Then the power $\mathfrak{m}^{\mathfrak{k}}$ shall be the product of the cardinal-number complex $\mathfrak{K}(\mathfrak{m})$; i. e., $\mathfrak{m}^{\mathfrak{k}} = \prod_{k \in \mathfrak{K}} \mathfrak{m}$. For $\mathfrak{m} \neq 0$, we put $\mathfrak{m}^0 = 1$.

In particular, according to this definition, $1^{\mathfrak{k}} = 1$ and $0^{\mathfrak{k}} = 0$ for $\mathfrak{k} \neq 0$.

It was shown on p. 38 that $\mathfrak{m}^{\mathfrak{k}}$ does not depend on the particular representative of the number $\mathfrak{m}$. That the power is also independent of the particular representative of the cardinal number $\mathfrak{k}$ follows directly from the commutative law on p. 38.

G. Cantor introduced the power in a somewhat different manner, which is independent of the general product. This definition is perhaps more transparent, and is particularly suited for many applications. We shall arrive at it here by specialization from our general definition.

According to our definition we may proceed as follows: Let $\mathfrak{M}(m)$ be a set having cardinal number $\mathfrak{m}$. We form new sets $\mathfrak{M}((k, m))$ from it by replacing, for an arbitrary, fixed element $k \in \mathfrak{K}$, every element m of $\mathfrak{M}$ by the pair of elements (k, m). The sets $\mathfrak{M}_k = \mathfrak{M}(k, m)$ thus obtained are mutually exclusive for the different k's, and their combinatorial product represents the power $\mathfrak{m}^{\mathfrak{k}}$. The combinatorial product is formed, however, in such a manner, that in $\mathfrak{K}(k)$ every element k is replaced by an arbitrary element of $\mathfrak{M}_k$, i. e., by an arbitrary pair of elements (k, m). Here the m's need by no means be different for different k's, because for different k's the dissimilarity of the pairs of elements is already taken care of by the prefixed k. The sets $\mathfrak{K}((k, m))$ arising in this way constitute the elements of the combinatorial product of the $\mathfrak{M}_k$'s. Every $\mathfrak{K}((k, m))$ can, however, also be described thus: With every element k of $\mathfrak{K}$ we associate an arbitrary element m of $\mathfrak{M}$. We have here a process which is analogous to the construction of a function $y = f(x)$; for, the function is also defined by a correspondence of certain values y to the values x. This association of elements m with the elements of $\mathfrak{K}$ Cantor calls a covering of the set $\mathfrak{K}$ with the set $\mathfrak{M}$. The set of all such coverings of the set $\mathfrak{K}$ with the set $\mathfrak{M}$ is called the covering set $\mathfrak{K} \mid \mathfrak{M}$. The power $\mathfrak{m}^{\mathfrak{k}}$ can accordingly be defined also as follows: Cover a set $\mathfrak{K}$ having cardinal number $\mathfrak{k}$, with a set $\mathfrak{M}$ having cardinal number $\mathfrak{m}$. The covering set $\mathfrak{K} \mid \mathfrak{M}$ then represents the cardinal number $\mathfrak{m}^{\mathfrak{k}}$.

For example, represent the cardinal numbers $\mathfrak{k} = 2$ and $\mathfrak{m} = 3$ by the sets $\mathfrak{K} = \{1, 2\}$ and $\mathfrak{M} = \{1, 2, 3\}$. Then the covering set is

$$\{\{(1, 1), (2, 1)\}, \{(1, 1), (2, 2)\}, \{(1, 1), (2, 3)\},$$

$$\{(1, 2), (2, 1)\}, \{(1, 2), (2, 2)\}, \{(1, 2), (2, 3)\},$$

$$\{(1, 3), (2, 1)\}, \{(1, 3), (2, 2)\}, \{(1, 3), (2, 3)\}\},$$

and it has cardinal number $9 = 3^2$. The point here is to cover the first place with one of the numbers 1, 2, 3, and, independently of this, to make an analogous covering of the second place. This gives 3^2 cases, as is easily seen even without actually executing the covering. In general, for finite cardinal numbers, $\mathfrak{m}^{\mathfrak{k}}$ yields the ordinary concept of power; of this one can convince oneself in the above manner, or it can be inferred from the fact that in this case the product of the cardinal numbers coincides exactly with the elementary product.

a) With the aid of the power concept, the *cardinal number of the set* $\mathfrak{U}(\mathfrak{N})$ *of all subsets* of a given set $\mathfrak{N}$ can now be given immediately; it is $|\mathfrak{U}(\mathfrak{N})| = 2^{\mathfrak{n}}$, where $\mathfrak{n} = |\mathfrak{N}|$.

For let $\mathfrak{N}_1$ be a given subset of $\mathfrak{N}$, and cover $\mathfrak{N}$ with the set $\{0, 1\}$ in such a manner, that with every element of $\mathfrak{N}$ is associated the number 0 or 1 according as this element does or does not appear in the subset $\mathfrak{N}_1$. The covering set $\mathfrak{N} \mid \{0, 1\}$ thus obtained is obviously equivalent to $\mathfrak{U}(\mathfrak{N})$, and has cardinal number $2^{\mathfrak{n}}$.

b) From a) and the theorem on p. 21 it follows at once that

$$2^{\mathfrak{n}} > \mathfrak{n},$$

and this also holds, trivially, for $\mathfrak{n} = 0$.

We shall now prove the customary rules of operation for powers.

c) $$\mathfrak{m}^{\mathfrak{k}_1} \cdot \mathfrak{m}^{\mathfrak{k}_2} = \mathfrak{m}^{\mathfrak{k}_1 + \mathfrak{k}_2} \quad \text{for } \mathfrak{k}_1, \mathfrak{k}_2 \neq 0.$$

Actually we shall need this rule in the following more general form:

c_1) If, with every element l of a set $\mathfrak{L}(l)$, there is associated a cardinal number $\mathfrak{k}_l \neq 0$, then for every cardinal number $\mathfrak{m}$ we have

$$(1) \qquad \prod_{l \in \mathfrak{L}} \mathfrak{m}^{\mathfrak{k}_l} = \mathfrak{m}^{\sum_{l \in \mathfrak{L}} \mathfrak{k}_l}.$$

This follows directly from Theorem 2 on p. 39, if, for the sets $\mathfrak{K}_l(k)$ there, we choose sets having cardinal number $\mathfrak{k}_l$, and put every $\mathfrak{m}_k = \mathfrak{m}$. For then, according to the definition of power, the left-hand side of equation (2) on p. 39 is just the left-hand side of (1), and the right-hand side of (2) is the right-hand side of (1).

d) $\mathfrak{m}_1^{\mathfrak{k}} \cdot \mathfrak{m}_2^{\mathfrak{k}} = (\mathfrak{m}_1 \cdot \mathfrak{m}_2)^{\mathfrak{k}}$ for $\mathfrak{k} \neq 0$.

This rule too follows from Theorem 2 on p. 39. For, first, we take $\mathfrak{L} = \{1, 2\}$, and for $\mathfrak{K}_1(k)$ and $\mathfrak{K}_2(k)$ we take two disjunct sets, each of cardinal number $\mathfrak{k}$, and associate with each k of $\mathfrak{K}_1$ the cardinal number $\mathfrak{m}_1$, and with each k of $\mathfrak{K}_2$ the cardinal number $\mathfrak{m}_2$. Then equation (2) on p. 39 states that

$$\mathfrak{m}_1^{\mathfrak{k}} \cdot \mathfrak{m}_2^{\mathfrak{k}} = \prod_{k \in \mathfrak{K}} \mathfrak{m}_k, \tag{2}$$

where $\mathfrak{K} = \mathfrak{K}_1 + \mathfrak{K}_2$. Next, on the basis of the relation $\mathfrak{K}_1 \sim \mathfrak{K}_2$, we choose a mapping of the sets $\mathfrak{K}_1$ and $\mathfrak{K}_2$ on each other. From each pair of corresponding elements of the two sets we form a set $\{k, k^*\}$, where $k \in \mathfrak{K}_1$ and $k^* \in \mathfrak{K}_2$. Then $\sum_{k \in \mathfrak{K}_1} \{k, k^*\} = \mathfrak{K}$. The set $\mathfrak{K}_1$ now plays the role of $\mathfrak{L}$ in Theorem 2, and every $\{k, k^*\}$ plays the role of a $\mathfrak{K}_l$. To every $k \in \mathfrak{K}_1$ corresponds the cardinal number $\mathfrak{m}_1$, and to every $k^* \in \mathfrak{K}_2$ corresponds the cardinal number $\mathfrak{m}_2$, so that equation (2) on p. 39 now states that

$$(\mathfrak{m}_1 \cdot \mathfrak{m}_2)^{\mathfrak{k}} = \prod_{k \in \mathfrak{K}} \mathfrak{m}_k. \tag{3}$$

The assertion d) is a consequence of (2) and (3).

e) $(\mathfrak{m}^{\mathfrak{k}})^{\mathfrak{l}} = \mathfrak{m}^{\mathfrak{k} \cdot \mathfrak{l}}$ for $\mathfrak{k}, \mathfrak{l} \neq 0$.

This follows immediately from c_1), if we take $\mathfrak{L}$ to be a set of cardinal number $\mathfrak{l}$ and put every $\mathfrak{k}_l = \mathfrak{k}$.

f) If $\mathfrak{k} \neq 0$, then $\mathfrak{m}_1 \leq \mathfrak{m}_2$ implies $\mathfrak{m}_1^{\mathfrak{k}} \leq \mathfrak{m}_2^{\mathfrak{k}}$.

This is a direct consequence of Theorem 3 on p. 40, since we have a product of $\mathfrak{k}$ factors on each side.

g) If $0 < \mathfrak{k}_1 \leq \mathfrak{k}_2$, then $\mathfrak{m}^{\mathfrak{k}_1} \leq \mathfrak{m}^{\mathfrak{k}_2}$.

For, let $\mathfrak{L}_1$ and $\mathfrak{L}_2$ be representatives of the cardinal numbers $\mathfrak{k}_1$ and $\mathfrak{k}_2$, and assume that $\mathfrak{m} \neq 0$. From the hypothesis, it follows that $\mathfrak{L}_1 \sim \overline{\mathfrak{L}}_2$ for a subset $\overline{\mathfrak{L}}_2 \subseteq \mathfrak{L}_2$. $\mathfrak{k}_1$ and $\mathfrak{k}_2$ can therefore be represented by two sets $\mathfrak{K}_1$ and $\mathfrak{K}_2$, where $\mathfrak{K}_1 \subseteq \mathfrak{K}_2$. If

$\mathfrak{K}_3 = \mathfrak{K}_2 - \mathfrak{K}_1$ has cardinal number $\mathfrak{k}_3$, then $\mathfrak{k}_1 + \mathfrak{k}_3 = \mathfrak{k}_2$. On account of f), $1^{\mathfrak{k}_3} \leq \mathfrak{m}^{\mathfrak{k}_3}$. Hence, multiplication by $\mathfrak{m}^{\mathfrak{k}_1}$ gives $\mathfrak{m}^{\mathfrak{k}_1} \leq \mathfrak{m}^{\mathfrak{k}_1 + \mathfrak{k}_3} = \mathfrak{m}^{\mathfrak{k}_2}$.

The result contained in b) is capable of a far-reaching generalization:

J. KÖNIG'S THEOREM. *To every element k of a given set $\mathfrak{K}(k)$, let there correspond two cardinal numbers $\mathfrak{m}_k$, $\mathfrak{n}_k$ such that invariably $\mathfrak{m}_k < \mathfrak{n}_k$. Then*

$$\sum_{k \in \mathfrak{K}} \mathfrak{m}_k < \prod_{k \in \mathfrak{K}} \mathfrak{n}_k .$$

Proof: Let the cardinal numbers $\mathfrak{n}_k$ be represented by the disjunct sets $\mathfrak{N}_k$. Since $\mathfrak{m}_k < \mathfrak{n}_k$, the $\mathfrak{m}_k$'s can be represented by subsets $\mathfrak{M}_k \subset \mathfrak{N}_k$. Finally, put $\mathfrak{R}_k = \mathfrak{N}_k - \mathfrak{M}_k$, and denote the elements of $\mathfrak{M}_k$, $\mathfrak{N}_k$, $\mathfrak{R}_k$ by m_k, n_k, r_k. Then

$$\sum_{k \in \mathfrak{K}} \mathfrak{m}_k = \Big| \sum_{k \in \mathfrak{K}} \mathfrak{M}_k \Big|, \qquad \prod_{k \in \mathfrak{K}} \mathfrak{n}_k = \Big| {}^{\times}\prod_{k \in \mathfrak{K}} \mathfrak{N}_k \Big|.$$

First we prove that $\sum \mathfrak{M}_k$ is equivalent to a subset of ${}^{\times}\prod = {}^{\times}\prod \mathfrak{N}_k$, from which it then follows that

$$\sum \mathfrak{m}_k \leq \prod \mathfrak{n}_k . \tag{4}$$

Let any element $m \in \sum \mathfrak{M}_k$ be given, and suppose that it belongs to $\mathfrak{M}_l$, say. With this element we are going to associate an element of ${}^{\times}\prod$. Every element of ${}^{\times}\prod$ is a set $\mathfrak{K}(n_k)$. We now choose a $\mathfrak{K}_m = \mathfrak{K}(n_k)$ in such a manner, that the element $n_l \in \mathfrak{N}_l$ appearing in $\mathfrak{K}$ is equal to m, and that n_k for $k \neq l$ is equal to an r_k. We let this $\mathfrak{K}_m$ correspond to m. Then, to different m's belong different $\mathfrak{K}_m$'s. Consequently, $\sum \mathfrak{M}_k$ is mapped on a subset of ${}^{\times}\prod$, and this proves (4).

Now we have to show that the equality sign in (4) does not hold. This is accomplished by showing that every set $\mathfrak{P} \subseteq {}^{\times}\prod$ which is equivalent to $\sum \mathfrak{M}_k$ is a *proper* subset of ${}^{\times}\prod$, and we prove this by means of a generalization of the second diagonal-method (cf. p. 9). Let $\mathfrak{P} \sim \sum \mathfrak{M}_k$ and $\mathfrak{P} \subseteq {}^{\times}\prod$. Then every $\mathfrak{M}_h$ is mapped on a part, $\mathfrak{P}_h$, of $\mathfrak{P}$, and $\sum_{h \in \mathfrak{K}} \mathfrak{P}_h = \mathfrak{P}$. Denote by $\mathfrak{K}_h(n_k)$ the elements belonging to $\mathfrak{P}_h$ ($\mathfrak{K}_h$

has nothing to do with the $\mathfrak{K}_m$ defined previously). $\mathfrak{M}_h \sim \mathfrak{P}_h$, and consequently to every $m_h \in \mathfrak{M}_h$ there corresponds exactly one $\mathfrak{K}_h \in \mathfrak{P}_h$ in a one-to-one manner, and hence also a "diagonal element" n_h of the set $\mathfrak{K}_h(n_k)$ in a unique (but perhaps no longer one-to-one) manner. Thus the set of distinct diagonal elements n_h which, for fixed h, can appear in the $\mathfrak{K}_h(n_k)$'s, has no greater cardinal number than $|\mathfrak{M}_h| = \mathfrak{m}_h < \mathfrak{n}_h$. Denote by $\mathfrak{N}_h^*$ the set of these different diagonal elements n_h. Then $\mathfrak{N}_h^* \subset \mathfrak{N}_h$, and hence $\mathfrak{R}_h^* = \mathfrak{N}_h - \mathfrak{N}_h^*$ is not empty. We now choose an arbitrary $r_h^* \in \mathfrak{R}_h^*$, and thereby obtain such an element r_h^* for every $h \in \mathfrak{K}$. With these elements we form the set $\mathfrak{K}(r_k^*)$. This set is an element of ${}^\times\prod$, but not of $\mathfrak{P}$. For, every element of $\mathfrak{P}$ belongs to exactly one $\mathfrak{P}_h$; and $\mathfrak{K}(r_k^*)$ differs from every $\mathfrak{K}_h(n_k) \in \mathfrak{P}_h$, because the elements with the index $h \in \mathfrak{K}$ in $\mathfrak{K}_h(n_k)$ and $\mathfrak{K}(r_k^*)$ are certainly distinct. Hence we have indeed $\mathfrak{P} \subset {}^\times\prod$, which completes the proof of the theorem.

If we choose, in particular, $\mathfrak{m} = 1$ and $\mathfrak{n} = 2$, then, for $\mathfrak{k} = |\mathfrak{K}|$, the theorem yields

$$\mathfrak{k} = \sum_{k \in \mathfrak{K}} 1 < \prod_{k \in \mathfrak{K}} 2 = 2^{\mathfrak{k}},$$

which is inequality b) once more.

If we choose for $\mathfrak{K}$ an enumerable set, and put

$$\mathfrak{s} = \mathfrak{m}_1 + \mathfrak{m}_2 + \mathfrak{m}_3 + \cdots, \quad \mathfrak{p} = \mathfrak{m}_1 \cdot \mathfrak{m}_2 \cdot \mathfrak{m}_3 \cdots$$

for

$$0 < \mathfrak{m}_1 < \mathfrak{m}_2 < \mathfrak{m}_3 < \cdots,$$

then the theorem gives

$$\begin{aligned} \mathfrak{m}_1 + \mathfrak{m}_2 + \mathfrak{m}_3 + \cdots &< \mathfrak{m}_2 \cdot \mathfrak{m}_3 \cdot \mathfrak{m}_4 \cdots \\ &= 1 \cdot \mathfrak{m}_2 \cdot \mathfrak{m}_3 \cdots \\ &\leq \mathfrak{m}_1 \cdot \mathfrak{m}_2 \cdot \mathfrak{m}_3 \cdots, \end{aligned}$$

and consequently

$$\mathfrak{s} < \mathfrak{p} = \mathfrak{m}_1 \cdot \mathfrak{m}_2 \cdot \mathfrak{m}_3 \cdots \leq \mathfrak{s}^{\mathfrak{s}}.$$

By exponentiation it follows from this that

$$\mathfrak{s}^{\mathfrak{a}} \leq \mathfrak{p}^{\mathfrak{a}} \leq \mathfrak{s}^{\mathfrak{a}\cdot\mathfrak{a}} = \mathfrak{s}^{\mathfrak{a}},$$

and hence

$$\mathfrak{s}^{\mathfrak{a}} = \mathfrak{p}^{\mathfrak{a}}.$$

11. Some Examples of the Evaluation of Powers

We shall first evaluate a few powers, and then interpret the results in various ways.

α) $m^n < \mathfrak{a}$, $\mathfrak{a}^n = \mathfrak{a}$, $\mathfrak{c}^n = \mathfrak{c}$ for every finite cardinal number $n \neq 0$. The first relation is trivial; the other two follow from the equations $\mathfrak{a}\cdot\mathfrak{a} = \mathfrak{a}$ and $\mathfrak{c}\cdot\mathfrak{c} = \mathfrak{c}$ obtained on pp. 30 and 31.

β) $$2^{\mathfrak{a}} = 10^{\mathfrak{a}} = n^{\mathfrak{a}} = \mathfrak{a}^{\mathfrak{a}} = \mathfrak{c}^{\mathfrak{a}} = \mathfrak{c} \qquad (n \geq 2).$$

The fact that $10^{\mathfrak{a}} = \mathfrak{c}$ follows from the definition of $\mathfrak{c}$ and from b) on p. 27. For, $10^{\mathfrak{a}}$ is represented by the set of all coverings of the enumerable set $\{1, 2, 3, \cdots\}$ with the numerals $0, 1, 2, \cdots, 9$. Each individual covering gives rise to an infinite decimal fraction $0\cdot a_1a_2a_3 \cdots$, and the set of all these infinite decimal fractions is obviously equivalent to the covering set. These infinite decimal fractions fall into two classes. The first class consists of those in which numerals different from 0 appear again and again; and the second class, of the rest. The first class is a set which is equivalent to the points of the interval $(0, 1\rangle$; and the second is enumerable, because it represents only rational numbers. Hence, $10^{\mathfrak{a}} = \mathfrak{c} + \mathfrak{a} = \mathfrak{c}$.

One can prove in a similar manner, that $2^{\mathfrak{a}} = \mathfrak{c}$ and $n^{\mathfrak{a}} = \mathfrak{c}$, by representing these powers by covering sets, and the individual coverings by dyadic or n-adic fractions. Suppose $2^{\mathfrak{a}} = \mathfrak{c}$, say, to be proved in that way. Then, as we shall see immediately, $n^{\mathfrak{a}} = \mathfrak{c}$ can be proved in yet another manner.

The fact that $\mathfrak{c}^{\mathfrak{a}} = \mathfrak{c}$ now results from $2^{\mathfrak{a}} = \mathfrak{c}$ according to the rules of operation as follows: We have

$$\mathfrak{c}^{\mathfrak{a}} = (2^{\mathfrak{a}})^{\mathfrak{a}} = 2^{\mathfrak{a}\cdot\mathfrak{a}} = 2^{\mathfrak{a}} = \mathfrak{c}.$$

From this we get, for $n \geq 2$,

$$\mathfrak{c} = 2^{\mathfrak{a}} \leq \begin{Bmatrix} n^{\mathfrak{a}} \\ \mathfrak{a}^{\mathfrak{a}} \end{Bmatrix} \leq \mathfrak{c}^{\mathfrak{a}} = \mathfrak{c}.$$

This proves all the relations in β).

From this it follows, e. g., that if in the sequence of cardinal numbers $\mathfrak{a}_1, \mathfrak{a}_2, \mathfrak{a}_3, \cdots$ we have invariably $2 \leq \mathfrak{a}_k \leq \mathfrak{c}$, then $\mathfrak{a}_1 \cdot \mathfrak{a}_2 \cdot \mathfrak{a}_3 \cdots = \mathfrak{c}$.

For,

$$\mathfrak{c} = 2^{\mathfrak{a}} \leq \mathfrak{a}_1 \cdot \mathfrak{a}_2 \cdot \mathfrak{a}_3 \cdots \leq \mathfrak{c}^{\mathfrak{a}} = \mathfrak{c}.$$

Hence, in particular,

$$\mathfrak{a}! = 1 \cdot 2 \cdot 3 \cdots = \mathfrak{c}.$$

Now we give, to begin with, two interpretations of the rules in α).

a) Let $\mathfrak{M}$ be a finite or an enumerable set. From the elements m of $\mathfrak{M}$, form, for a fixed number n, *all* n-termed complexes $(m_1, m_2, \cdots, m_n)$. Here the m_k's need by no means be *distinct* elements of $\mathfrak{M}$. We thus obtain the coverings of the set $\{1, 2, \cdots, n\}$ with the set $\mathfrak{M}$. This covering set has cardinal number $|\mathfrak{M}|^n \leq \mathfrak{a}$. The set of *all* n-termed complexes for $n = 1, 2, 3, \cdots$ then has cardinal number $\sum_n |\mathfrak{M}|^n \leq \mathfrak{a} \cdot \mathfrak{a} = \mathfrak{a}$.

If, in particular, we take $\mathfrak{M}$ to be an "alphabet", then we can say (Hausdorff [1], p. 61): "From an 'alphabet', i. e., a finite set of 'letters', we can form an enumerable set of finite complexes of letters, i. e., 'words', some of which, of course, like abracadabra, are meaningless. If further elements, such as punctuation marks, printing spaces, numerals, notes, etc., are added to these letters, it is clear that the set of all books, catalogues, symphonies, operas, is also enumerable, and would remain enumerable if we were to employ even an enumerable number of symbols (but for each complex only a finite number). On the other hand, if we have a finite number of symbols, and we limit the complexes to a certain maximum number of elements by declaring words of more than one hundred letters and books of more than one million words inadmissible, then these sets become finite. And if we assume, with Giordano Bruno, the existence of an infinite number of heavenly bodies with speaking, writing, and music-making inhabitants, then it follows with mathematical certainty, that on infinitely many of these heavenly bodies the same opera must be performed, with the same libretto, the same names of the composer, the librettist, the members of the orchestra, and the singers."

b) Space of n dimensions contains "just as many" points as the segment $\langle 0, 1 \rangle$. For, every point in n-dimensional space is determined by a complex of n numbers $x_1, x_2, \cdots, x_n$. Each of these number complexes is a covering of the numbers 1, $2, \cdots, n$ with real numbers. The set of all these coverings has cardinal number $\mathfrak{c}^n = \mathfrak{c}$.

From the results in β) we can infer the following:

c) Between 0 and 1 there are "just as many" dyadic fractions as there are decimal fractions or n-adic fractions, and the set of all dyadic fractions between 0 and 1 comprises "just as many" elements as the set of *all* dyadic fractions. This follows from the fact that the set of n-adic fractions is always equivalent to the set of all coverings of an enumerable set with the numbers $1, 2, \cdots, n$, whether we consider all of these fractions or only those lying between 0 and 1; and that this covering set has cardinal number $n^{\mathfrak{a}} = \mathfrak{c}$, so that its cardinal number is independent of n.

d) A point in space of enumerably many dimensions is understood to be a system of enumerably many numbers $x_1, x_2, x_3, \cdots$. The totality of all points in space of enumerably many dimensions is "just as large" a set as the set of all points of the interval $\langle 0, 1 \rangle$. For, the first set is the set of all coverings of an enumerable set with the set of all real numbers, and it therefore has power $\mathfrak{c}^{\mathfrak{a}} = \mathfrak{c}$.

e) If we understand a *lattice point* in space of enumerably many dimensions to be a system of *whole* numbers $x_1, x_2, x_3, \cdots$, then we can say that the "number" of all lattice points in space of enumerably many dimensions "coincides" with the "number" of *all* points in space of enumerably many dimensions. For, the first set is the set of all coverings of an enumerable set with an enumerable set, and it therefore has cardinal number $\mathfrak{a}^{\mathfrak{a}} = \mathfrak{c}$.

f) The set of all real, continuous functions has the same power as the set of all real, analytic functions, be these functions defined only in an interval or for all values of the argument; each set has power $\mathfrak{c}$ (Borel).

For, each of the sets contains, in particular, the constant functions. Their power, however, is $\mathfrak{c}$, since they can assume an arbitrary real value. Therefore the power of each of the two sets of functions is at least $\mathfrak{c}$. To obtain an upper estimate, we need consider only the continuous functions, because the class of analytic functions is contained in the class of continuous functions. Now a continuous function is already known if it is known at merely an everywhere dense set of points[7] (where, however, it cannot, of course, be defined completely arbitrarily), say at the rational points. But the set of all functions defined at the rational points can be regarded as the set of all coverings of the rational numbers with the real numbers, and this covering set has cardinal number $\mathfrak{c}^{\mathfrak{a}} = \mathfrak{c}$. The cardinal number of each of our sets is thus $\geq \mathfrak{c}$ and $\leq \mathfrak{c}$, and is therefore equal to $\mathfrak{c}$.

Remark. The same result is obtained if the functions are allowed to assume only the values of a certain interval.

Concerning powers with exponent $\mathfrak{c}$, we have

$$\gamma) \qquad 2^{\mathfrak{c}} = n^{\mathfrak{c}} = \mathfrak{a}^{\mathfrak{c}} = \mathfrak{c}^{\mathfrak{c}} = \mathfrak{f} \qquad (n \geq 2).$$

For, $\mathfrak{f}$ was defined as the cardinal number of the set of all real functions in the interval $\langle 0, 1\rangle$. This set, however, is the set of all coverings of the continuum with the set of all real numbers, so that its cardinal number can also be represented by $\mathfrak{c}^{\mathfrak{c}}$. This proves the last of the above relations. From this follows

$$2^{\mathfrak{c}} = 2^{\mathfrak{a}\cdot\mathfrak{c}} = (2^{\mathfrak{a}})^{\mathfrak{c}} = \mathfrak{c}^{\mathfrak{c}} = \mathfrak{f},$$

and hence

$$\mathfrak{f} = 2^{\mathfrak{c}} \leq \begin{Bmatrix} n^{\mathfrak{c}} \\ \mathfrak{a}^{\mathfrak{c}} \end{Bmatrix} \leq \mathfrak{c}^{\mathfrak{c}} = \mathfrak{f},$$

which proves all the equations in γ).

---

[7] A set of points is said to be *everywhere dense* on a straight line, if every subinterval contains at least one point of the set.

Now we infer, e. g.:

g) The set of all real functions which assume only the values 0 and 1, has the same cardinal number as the set of all real functions of one variable. For, the first set of functions has cardinal number $2^{\mathfrak{c}} = \mathfrak{f}$.

h) The set of all real functions of enumerably many variables has the same cardinal number as the set of real functions of a single variable. For, the points of space of enumerably many dimensions constitute a set of cardinal number $\mathfrak{c}$. Therefore the set of all real functions of enumerably many variables has cardinal number $\mathfrak{c}^{\mathfrak{c}} = \mathfrak{f}$.

As a simple application of the rules of operation, we note, in addition, that

$$\mathfrak{f}^{n} = \mathfrak{f}^{\mathfrak{a}} = \mathfrak{f}^{\mathfrak{c}} = \mathfrak{f} \qquad (n \geq 1).$$

CHAPTER III

Ordered Sets and Their Order Types

1. Definition of Ordered Set

In the preceding chapter, the concept of "number" was extended, and in such a manner, that transfinite numbers were added to the finite ones. These numbers, finite and transfinite, were called cardinal numbers. Now already in connection with the ordinary process of counting finite sets, every number plays a double role. Grammar, too, distinguishes numbers into cardinal numbers (fundamental numbers) and ordinal numbers (order numbers). For with finite sets, numbers can also be used to determine a certain order of succession of the elements. For instance, one can stipulate that in the set $\{a, b, c\}$, the first element shall be c, the second, a, and the third, b. It is then convenient to write the set in the form $\{c, a, b\}$. Which of two distinct elements precedes the other is hereby determined. Such determinations will now be carried over to arbitrary sets.

Sets for whose elements a certain order of succession has been prescribed play an important role in applications. If, e. g., a set of real numbers is given, then of every pair of distinct numbers of this set, one is the smaller of the two. Any set of real numbers can thus be ordered in an entirely natural fashion according to the magnitude of its elements. An order for the points on a segment suggests itself in just as natural a manner.

Since the elements of sets need by no means be numbers, or points of a segment, and; furthermore, even if the elements are numbers, it is not at all necessary for them to be ordered according to magnitude (e. g., the order of succession $\{3, 1, 2\}$ can be prescribed), the concept of ordered set must be formulated quite abstractly. We thus arrive at

Definition 1. *A set $\mathfrak{M}$ is called an ordered set, provided that a relation, denoted, say, by the symbol $\prec$, subsists between every*

pair of distinct elements a and b of the set, and only between dis-stinct elements, and satisfies the following two conditions:

1. *If $a \neq b$, then either $a \prec b$ or $b \prec a$.*

2. *If $a \prec b$ and $b \prec c$, then invariably $a \prec c$; i. e., the relation is transitive.*

We shall regard the empty set, as well as a set consisting of a single element, as ordered; likewise, a set consisting of two elements, provided that an ordering relation is given for its two elements a, b, according to which precisely one of the two relations $a \prec b$, $b \prec a$ holds.

The relation $a \succ b$ shall mean the same as $b \prec a$. Finally, the symbol $a \prec b$ is read "a precedes b"; and the symbol $a \succ b$, "a succeeds b".

This last terminology is by no means intended to give the symbol an intuitive meaning; it serves merely to facilitate comprehension. Moreover, the meaning of the symbol $\prec$ may change from set to set. For instance, in the case of sets of numbers, it may coincide at times with the symbol $<$, and at other times, with the symbol $>$. A few examples will serve to clarify the concept of ordered set.

a) The set of natural numbers $\{1, 2, 3, \cdots\}$, ordered according to increasing magnitude. Here $\prec$ coincides with $<$.

b) The set of natural numbers, ordered according to decreasing magnitude. We shall indicate this order by writing $\{\cdots, 3, 2, 1\}$. Here $\prec$ coincides with $>$.

c) The set of all integers, ordered according to increasing magnitude. This order is symbolized by

$$\{\cdots, -3, -2, -1, 0, 1, 2, 3, \cdots\}.$$

Here $\prec$ coincides with $<$.

d) The set of integers, but this time in the order indicated by $\{0, 1, -1, 2, -2, \cdots\}$; i. e., the nonnegative integers are ordered according to increasing magnitude, and each negative integer comes immediately after the positive integer having the same absolute value. The symbol $\prec$ in this case coincides in part with $<$, in part with $>$.

e) $\{0, 2, 4, \cdots, 1, 3, 5, \cdots\}$; i. e., first come the even, nonnegative integers ordered according to increasing magnitude, and then the odd integers.

f) $\{0, 2, 4, \cdots, 5, 3, 1\}$; i. e., first come the even, nonnegative integers ordered according to increasing magnitude, then the odd integers ordered according to decreasing magnitude.

g) $\{1, 2, 3, \cdots, \frac{1}{2}, \frac{3}{2}, \frac{5}{2}, \cdots, \frac{1}{3}, \frac{2}{3}, \frac{4}{3}, \cdots, \cdots\}$; i. e., the positive rational numbers ordered according to increasing denominator, and, for the same denominator, according to increasing numerator.

h) The set of all real numbers in the interval (0, 1) ordered according to decreasing magnitude.

The question whether every given set can be ordered will not be decided until later (p. 110). It is clear that every finite set can be ordered.

We have obviously

THEOREM 1. *Let $\mathfrak{M}$ be an ordered set. Then, by means of the ordering principle given for $\mathfrak{M}$, every subset of $\mathfrak{M}$ is also ordered. Hence, the subsets of ordered sets can always be regarded as ordered.*

Definition 2. *Two ordered sets $\mathfrak{M}$ and $\mathfrak{N}$ are said to be equal, in symbols: $\mathfrak{M} = \mathfrak{N}$, if they contain the same elements, and if the ordering relation $a \prec b$ valid for $\mathfrak{M}$ invariably implies the relation $a \,{}^*\!\!\prec b$ valid for $\mathfrak{N}$. Conversely, then, it follows obviously that $a \,{}^*\!\!\prec b$ always implies $a \prec b$.*

Thus, the sets a) and b) are to be considered as distinct in the sense of ordered sets. Without regard to an order, however, they are equal.

Finally, we introduce the following terms: If, for three elements a, b, c of a set, we have $a \prec b \prec c$, then we shall say that b lies between a and c. If an ordered set $\mathfrak{M}$ contains an element a such that for every $b \in \mathfrak{M}$ with $b \neq a$ the relation $a \prec b$ holds, then a will be called a first element of the set $\mathfrak{M}$. If $\mathfrak{M}$ contains an element c such that for every $b \in \mathfrak{M}$ with $b \neq c$ the relation $c \succ b$ holds, then c will be called a last element of $\mathfrak{M}$.

THEOREM 2. *An ordered set contains at most one first element and at most one last element.*

Proof: If the set contained, e. g., two first elements a and $\bar{a}$, we should have $a \prec \bar{a}$ as well as $\bar{a} \prec a$, which is impossible.

An ordered set, however, need have neither a first nor a last element. Examples of this are the sets c) and h).

2. *Similarity and Order Type*

For ordered sets, those mappings of one set on another will, naturally, be of importance, which, to put it briefly, preserve the order of succession of the elements. In this way we arrive at the following sharpening of the equivalence concept:

Definition 1. *An ordered set* $\mathfrak{M}$ *with the ordering relation* $\prec$ *is said to be similar to an ordered set* $\mathfrak{N}$ *with the ordering relation* $^*\prec$, *in symbols:* $\mathfrak{M} \simeq \mathfrak{N}$, *if the set* $\mathfrak{M}$ *can be mapped on the set* $\mathfrak{N}$ *in such a manner, that if* m_1 , m_2 *are any two elements of* $\mathfrak{M}$, *and* n_1 , n_2 *are the corresponding elements of* $\mathfrak{N}$, *then* $m_1 \prec m_2$ *always implies* $n_1 \ {}^*\!\prec n_2$. *It follows then obviously that* $n_1 \ {}^*\!\prec n_2$ *always implies that, for the corresponding elements,* $m_1 \prec m_2$. *Such a mapping is called a similarity mapping.*

The four fundamental properties result immediately from the definition:

α) $\mathfrak{M} \simeq \mathfrak{M}$, i. e., every ordered set is similar to itself.

β) $\mathfrak{M} \simeq \mathfrak{N}$ implies $\mathfrak{N} \simeq \mathfrak{M}$.

γ) If $\mathfrak{M} \simeq \mathfrak{N}$ and $\mathfrak{N} \simeq \mathfrak{P}$, then $\mathfrak{M} \simeq \mathfrak{P}$.

δ) $\mathfrak{M} \simeq \mathfrak{N}$ implies $\mathfrak{M} \sim \mathfrak{N}$.

Let us first clarify the concept of similarity by a few examples.

a) If an ordered set is mapped on itself, it is by no means necessary that every element correspond invariably to itself. If, e. g., $\mathfrak{M}$ is the set of numbers in the interval (0, 1) ordered according to increasing magnitude, then the function $y = x^2$ maps this interval on itself. No point corresponds to itself, however, under this similarity mapping.

b) An ordered set can be similar to a proper subset. For example, $\{1, 2, 3, \cdots\} \simeq \{2, 3, 4, \cdots\}$, and the set $\{1, 2, 3, \cdots\}$ is similar to the set of prime numbers if these are ordered according to increasing magnitude, because there are infinitely many prime numbers.

c) The set of rational numbers can be ordered in such a

manner, that it is similar to the ordered set $\{1, 2, 3, \cdots\}$. This is accomplished by the order of the rational numbers defined on p. 3. More generally, we have

THEOREM 1. *If a set* $\mathfrak{N}$ *is equivalent to an ordered set* $\mathfrak{M}$, *then* $\mathfrak{N}$ *can be ordered in such a way that* $\mathfrak{N} \simeq \mathfrak{M}$.

Proof: If m_1, m_2 are two elements of $\mathfrak{M}$, and if $m_1 \prec m_2$, then, for the corresponding elements n_1, n_2 of $\mathfrak{N}$, we stipulate that $n_1 \prec n_2$. This defines an ordering relation for the elements of $\mathfrak{N}$; and the relation is transitive within the set $\mathfrak{N}$, because, by hypothesis, it possesses this property in the set $\mathfrak{M}$. According to Definition 1 on p. 52, the set $\mathfrak{N}$ is ordered by this relation.

The following theorem is sometimes useful for investigating the similarity of given sets.

THEOREM 2. *If two sets are similar, then either* both *possess first (last) elements or* neither *of the two sets has such an element.*

Proof: If one of the sets has a first element m_0, then for every other element m of this set we have $m_0 \prec m$. Let n_0 denote that element which corresponds to m_0 under the similarity mapping of the one set on the other, and likewise let n be the element corresponding to m. Then, according to Definition 1, $n_0 \prec n$ for every $n \neq n_0$; i. e., the set $\mathfrak{N}$ also has a first element.

d) The sets a) and b) on p. 53 are not similar, because a) has a first element and b) does not. Further, with the aid of Theorem 2 follows, e. g., the dissimilarity of the sets a) and c), b) and c), a) and f), b) and f), h) and f), h) and a).

e) If we consider the numbers of the two intervals $\langle 0, 1\rangle$ and $(0, 1)$, ordered in each case according to increasing magnitude, then the two sets are dissimilar. Consequently, under a mapping of the two equivalent sets on one another, the order of succession of the elements in one of the two sets must of necessity be disturbed, as, for example, on p. 15.

f) If $\mathfrak{M}$ is the set of real numbers in the interval $(0, 1)$, and $\mathfrak{N}$ is the set of *all* real numbers, each set ordered according to increasing magnitude of its elements, then $\mathfrak{M} \simeq \mathfrak{N}$. This follows from the mapping given in c) on p. 15, which is evidently a similarity mapping.

g) Let $\mathfrak{M}$ be the set of all points $P(x, y)$ in the plane with rectangular coordinates x, y. Order this set so that

$$P(x_1, y_1) \prec P(x_2, y_2) \text{ for } x_1 < x_2 \text{ as well as for } x_1 = x_2, y_1 < y_2 .$$

This relation is obviously transitive, so that it actually defines an order for the set $\mathfrak{M}$. Further, let $\mathfrak{N}$ be the set of points $P(x, y)$ of the square $0 < x < 1$, $0 < y < 1$, ordered according to the same rules. Then $\mathfrak{M} \simeq \mathfrak{N}$. This is seen immediately by mapping the sides of the square similarly on the complete coordinate axes, in virtue of f). It will turn out later (p. 75), by the way, that the set $\mathfrak{M}$ is not similar to the interval $(0, 1)$.

With the aid of the concept of similarity, we can now once more introduce a new kind of number according to the pattern of ch. II, §3:

Definition 2. *By an order type* μ *we mean an arbitrary representative* $\mathfrak{M}$ *of a class of mutually similar ordered sets. The order type of an ordered set* $\mathfrak{M}$ *will sometimes be denoted by* $\overline{\mathfrak{M}}$. *The notation* $\overline{\mathfrak{M}}$ *accordingly means simply that the ordered set* $\mathfrak{M}$ *may be replaced by any ordered set similar to it.*

Since all sets that are similar to each other are also equivalent to each other, all sets possessing the same order type have the same cardinal number too; i. e., $\overline{\mathfrak{M}} = \overline{\mathfrak{N}}$ always implies $|\mathfrak{M}| = |\mathfrak{N}|$. Consequently, to every order type μ belongs a cardinal number, which we shall denote by $|\mu|$. Thus, $\mu = \nu$ implies $|\mu| = |\nu|$.

If two finite sets are equivalent, then it is obvious that, no matter how their elements are ordered, they are always similar too. Hence, to every finite cardinal number there corresponds precisely one order type, which, for brevity, will be denoted by the appropriate cardinal number. In the case of finite sets, then, cardinal number and order type coincide.

The order type of the set of natural numbers, ordered according to increasing magnitude, is denoted by ω:

$$\omega = \overline{\{1, 2, 3, \cdots\}}.$$

If, on the other hand, these numbers are ordered according to decreasing magnitude, then the order type of the set is denoted by $^*\omega$:

$$^*\omega = {}_{|}\{\cdots, 3, 2, 1\}^{|}.$$

In general, $^*\mu$ denotes that order type which results from μ when the order of succession of the elements is reversed, i. e., when a new ordering relation is derived from the relation $\prec$ defined for the original set, by setting $^*\prec$ equal to $\succ$.

We speak here not of order *numbers*, but of order *types*, and for the following reason. If we wish to compare two distinct order types, we must say, corresponding to the comparison of cardinal numbers introduced on p. 18, that the order type μ is smaller than the order type ν if a representative $\mathfrak{M}$ of μ is similar to a subset of a representative $\mathfrak{N}$ of ν. Let us now consider the order types ω and $^*\omega$. Every subset of the set $\{\cdots, 3, 2, 1\}$ which represents the order type $^*\omega$, has a last element, and therefore cannot be similar to the set $\{1, 2, 3, \cdots\}$ which represents the order type ω. Likewise we see that the first set cannot be similar to any subset of the second either. These order types are therefore not comparable. Order types thus lack an important property possessed by numbers. For this reason the term order *number* is not applied at this point.

3. *The Sum of Order Types*

For the purpose of defining addition of order types, we shall first introduce an ordered addition of ordered sets.

Definition 1. *Let $\mathfrak{M}$ and $\mathfrak{N}$ be two disjunct, ordered sets. We define an ordering relation for the elements s of their union $\mathfrak{S}$ as follows: Let s_1 and s_2 be two elements of the union. If they both belong to $\mathfrak{M}$, we let $s_1 \prec s_2$ or $s_2 \prec s_1$ in the union, according as the first or second of these relations holds for these elements in $\mathfrak{M}$. If both elements belong to $\mathfrak{N}$, then the relation which is valid for them in $\mathfrak{N}$ shall again hold also in the union. But if one of the elements belongs to $\mathfrak{M}$, and the other belongs to $\mathfrak{N}$, say $s_1 \in \mathfrak{M}$, $s_2 \in \mathfrak{N}$, then we let $s_1 \prec s_2$ in the union. The ordering relation thus defined for the union is obviously transitive, and it therefore* *orders $\mathfrak{S}$. The ordered sum $\mathfrak{M} + \mathfrak{N}$ of the ordered sets $\mathfrak{M}$ and $\mathfrak{N}$ is now understood to be the union $\mathfrak{S}$, ordered in the manner indicated. We speak here of an ordered addition of the sets $\mathfrak{M}$ and $\mathfrak{N}$. It is obvious that, when dealing with ordered addition, the sum $\mathfrak{M} + \mathfrak{N}$ is to be distinguished from the sum $\mathfrak{N} + \mathfrak{M}$.*

For example, if the sums are ordered sums,

$$\{1, 2, 3\} + \{4, 5, 6, 7\} = \{1, 2, 3, 4, 5, 6, 7\},$$

but

$$\{4, 5, 6, 7\} + \{1, 2, 3\} = \{4, 5, 6, 7, 1, 2, 3\}.$$

Definition 2. *To obtain the sum of the two order types μ and ν, represent them by two disjunct sets $\mathfrak{M}$ and $\mathfrak{N}$ and form the ordered sum $\mathfrak{S} = \mathfrak{M} + \mathfrak{N}$. Then we stipulate that $\mu + \nu = {}_{|}\mathfrak{S}^{|}$.*

It is again easy to see that the sum $\mu + \nu$ is independent of the particular representatives of the two order types. Further, from the definition of sum, it follows that

$$\alpha) \qquad |\mu + \nu| = |\mu| + |\nu|.$$

It is clear that the commutative law holds for the addition of finite order-types. *For transfinite order-types, however, addition need not be commutative.* For if

$$\mathfrak{M} = \{1, 2, 3, \cdots\} \quad \text{and} \quad \mathfrak{N} = \{0\},$$

then

$$\mathfrak{N} + \mathfrak{M} = \{0, 1, 2, 3, \cdots\}, \quad \mathfrak{M} + \mathfrak{N} = \{1, 2, 3, \cdots, 0\},$$

so that $1 + \omega = \omega$, but $\omega + 1 \neq \omega$, because the order type $\omega + 1$ has a last element, whereas ω does not.

We take a few more examples to illustrate addition of order types:

$$\text{a)} \qquad n + \omega = \omega.$$

For we have

$$n = {}_{|}\{1, 2, \cdots, n\}^{|}, \quad \omega = {}_{|}\{n + 1, n + 2, n + 3, \cdots\}^{|},$$

and hence

$$n + \omega = {}_{|}\{1, 2, \cdots, n, n + 1, \cdots\}^{|} = \omega.$$

b) $$^*\omega + n = {}^*\omega.$$

For,

$$^*\omega + n = {}_{|}\{\cdots, n+2, n+1\}' + {}_{|}\{n, \cdots, 2, 1\}'$$
$$= {}_{|}\{\cdots, n+1, n, \cdots, 1\}'.$$

c) $\omega, \omega + 1, \omega + 2, \cdots$ are distinct order types. For if we had $\omega + m = \omega + n$, with, say, $0 \leq m < n$, then $\omega + m$ would have to have an element that is followed by precisely $n - 1$ elements, which is not the case.

d) One shows in a similar manner that $^*\omega$, $1 + {}^*\omega$, $2 + {}^*\omega$, $\cdots$ are distinct order types.

e) The order type $^*\omega + \omega$ belongs, e. g., to the ordered set $\{\cdots, -3, -2, -1, 0, 1, 2, 3, \cdots\}$; the order type $\omega + {}^*\omega$, to the set $\{1, 2, 3, \cdots, -3, -2, -1\}$.

Although the commutative law ordinarily does not hold for order types, at least the associative law is valid:

β) $$(\mu + \nu) + \pi = \mu + (\nu + \pi).$$

This follows directly from the definition of addition. For if the order types are represented by the mutually exclusive sets $\mathfrak{M}$, $\mathfrak{N}$, $\mathfrak{P}$, then for the ordered addition of these sets we have obviously $(\mathfrak{M} + \mathfrak{N}) + \mathfrak{P} = \mathfrak{M} + (\mathfrak{N} + \mathfrak{P})$.

From this we get, using the results obtained previously:

f) $$n + \omega + \omega = (n + \omega) + \omega = \omega + \omega;$$

$$\omega + n + \omega = \omega + (n + \omega) = \omega + \omega;$$

$$\omega + \omega + n = \omega + \omega + n.$$

g) $$^*\omega + n + \omega = ({}^*\omega + n) + \omega = {}^*\omega + \omega;$$

$$n + {}^*\omega + \omega \quad \text{and} \quad {}^*\omega + \omega + n$$

are distinct order types.

Addition can also be extended to arbitrarily many order types. We begin with

Definition 3. *Let an ordered set* $\mathfrak{K}(k)$ *be given, and to every element k let there correspond an ordered set* $\mathfrak{M}_k$, *where the* $\mathfrak{M}_k$*'s are mutually exclusive. Then we order the union* $\mathfrak{S} = \sum_{k \in \mathfrak{K}} \mathfrak{M}_k$ *in the following manner: If two elements* s_1 *and* s_2 *of* $\mathfrak{S}$ *belong to the same* $\mathfrak{M}_k$, *then we let* $s_1 \prec s_2$ *or* $s_2 \prec s_1$ *in* $\mathfrak{S}$, *according as the first or second relation is valid for these elements in* $\mathfrak{M}_k$. *If, however,* s_1 *and* s_2 *belong to different* $\mathfrak{M}_k$*'s, say* $s_1 \in \mathfrak{M}_{k_1}$, $s_2 \in \mathfrak{M}_{k_2}$, *then we let* $s_1 \prec s_2$ *or* $s_2 \prec s_1$ *in* $\mathfrak{S}$, *according as* $k_1 \prec k_2$ *or* $k_2 \prec k_1$. *The ordered sum* $\sum_{k \in \mathfrak{K}} \mathfrak{M}_k$, *now, shall equal the union* $\mathfrak{S}$, *ordered as indicated.*

Definition 4. *Let an ordered set* $\mathfrak{K}(k)$ *be given, and to every element k of this set let there correspond an order type* μ_k; *in other words, let there be given an ordered complex of order types, arising from the ordered set* $\mathfrak{K}(k)$. *Represent the order types* μ_k *by mutually exclusive ordered sets* $\mathfrak{M}_k$. *Then the sum of this complex of order types shall be*

$$\sum_{k \in \mathfrak{K}} \mu_k = \left| \sum_{k \in \mathfrak{K}} \mathfrak{M}_k \right|,$$

where the addition of the sets on the right-hand side is to be carried out as an ordered addition.

It is easy to see once more that this sum is independent of the particular representatives of the order types. Further, we have obviously:

$$\gamma) \qquad \left| \sum_{k \in \mathfrak{K}} \mu_k \right| = \sum_{k \in \mathfrak{K}} |\mu_k|.$$

Here, too, of course, the universal validity of the commutative law is out of the question. It is true that there is an associative law here which goes beyond β), but we shall not set it up, because it is not needed in what follows.

4. *The Product of Two Order Types*

If, in particular, the same order type μ is associated with every k in §3, Definition 4, we arrive at

Definition 1. *Let there be given two order types* $\kappa \neq 0$ *and* μ. *Represent the order type* κ *by the set* $\mathfrak{K}(k)$, *and with every k associate*

the order type μ. *Then the product* $\mu \cdot \kappa$ *shall equal the sum of the complex of order types thus determined; i. e.,*

$$\mu \cdot \kappa = \sum_{k \in \mathfrak{K}} \mu.$$

For $\kappa = 0$ *we put* $\mu \cdot \kappa = 0$.

That the product is independent of the particular representative of μ was already stated on p. 61. It is also easy to see that it is independent of the particular representative of κ too; this will, moreover, follow once more, presently. The product can also be defined as follows:

Definition 2. *Let there be given two order types* $\kappa \neq 0$ *and* $\mu \neq 0$. *Represent* κ *by* $\mathfrak{K}(k)$ *and* μ *by* $\mathfrak{M}(m)$. *Form the set of all pairs of elements* (k, m), *and define an order for these pairs by stipulating that*

$(k_1, m_1) \prec (k_2, m_2)$ *for* $k_1 \prec k_2$ *as well as for* $k_1 = k_2$, $m_1 \prec m_2$. *Let* $\mathfrak{S}$ *denote the set of pairs of elements* (k, m) *ordered in this manner.* $\mathfrak{S}$ *will also be called the ordered product of the second kind,* $\mathfrak{M} \times \mathfrak{K}$ *(note the order in which the factors are written down). Then the product* $\mu \cdot \kappa$ *shall equal* $|\mathfrak{S}|$. *If* $\kappa = 0$ *or* $\mu = 0$, *we put* $\mu \cdot \kappa = 0$. *The names lexicographical order and order according to first differences readily suggest themselves for the order, just described, of the pairs of elements* (k, m).

This second definition follows immediately from the first. The set $\mathfrak{K}$ is the same in both. In the first definition we associate with every $k \in \mathfrak{K}$ a set $\mathfrak{M}_k$ with order type μ, and all the $\mathfrak{M}_k$'s must be mutually exclusive. This can be done by taking a set $\mathfrak{M}$ with order type μ, and replacing the elements m of $\mathfrak{M}$, for every k, by (k, m). If we now associate with every k the set $\mathfrak{M}((k, m))$, the requirement of Definition 1 is fulfilled. The union of the $\mathfrak{M}((k, m))$'s is precisely the set denoted by $\mathfrak{S}$ in Definition 2, and the order prescribed for its elements in Definition 2 is also that prescribed in Definition 1.

It is immediately evident from Definition 2, that the product of the order types is independent of their particular representatives. Further, it follows directly from each of the two definitions, that

$$\alpha) \qquad |\mu\cdot\kappa| = |\mu|\cdot|\kappa|.$$

For multiplication there is no commutative law either. For if we first choose $\kappa = \omega$, $\mu = 2$, then $2\cdot\omega$ is represented, e. g., by the ordered set $\{a_1, b_1, a_2, b_2, a_3, b_3, \cdots\}$, and is therefore equal to ω. If we then choose $\kappa = 2$, $\mu = \omega$, then $\omega\cdot 2$ is represented, e. g., by $\{1, 3, 5, \cdots; 2, 4, 6, \cdots\}$. This set is certainly not similar to the first one, because the second contains elements (e. g., the element 2) which are preceded by infinitely many elements, whereas no such elements appear in the first set. Hence, $\omega\cdot 2 \neq 2\cdot\omega$. With products too, then, one must pay attention to the order of succession of the factors. Under these circumstances it would undoubtedly be more agreeable if the product defined above were denoted by $\kappa\cdot\mu$ instead of $\mu\cdot\kappa$. Nevertheless, the notation introduced above has become accepted, and a change of notation at this time would lead to great confusion in the literature. Consequently, one must always bear in mind that when considering the product $\mu\cdot\kappa$, the number pairs (k, m), and not (m, k), are to be ordered according to first differences.

We do have, however, for multiplication, as for addition, the associative law, which is so important for calculation:

$$\beta) \qquad (\mu\cdot\nu)\cdot\pi = \mu\cdot(\nu\cdot\pi).$$

For let the order types be represented by sets with the elements m, n, p. Then, according to Definition 2, the order type $(\mu\cdot\nu)\cdot\pi$ is represented by the set of elements $(p, (n, m))$, where

$$(p_1, (n_1, m_1)) \prec (p_2, (n_2, m_2))$$

if $p_1 \prec p_2$, or $p_1 = p_2$, $n_1 \prec n_2$, or $p_1 = p_2$, $n_1 = n_2$, $m_1 \prec m_2$. Likewise, $\mu\cdot(\nu\cdot\pi)$ is represented by the set of elements $((p, n), m)$, where

$$((p_1, n_1), m_1) \prec ((p_2, n_2), m_2)$$

once more under the conditions just cited. If we now associate, for the same elements m, n, p, the element $(p, (n, m))$ with the element $((p, n), m)$, this defines a similarity correspondence between the sets representing $(\mu\cdot\nu)\pi$ and $\mu(\nu\cdot\pi)$, which proves the assertion.

The distributive law does not always hold. It does not always hold, namely, when the *first* factor is a sum. For example,

$$(\omega + 1)\cdot 2 = \omega + 1 + \omega + 1 = \omega + \omega + 1 = \omega\cdot 2 + 1,$$

and this differs from

$$\omega\cdot 2 + 1\cdot 2 = \omega\cdot 2 + 2,$$

because the last order type has a next to the last element, whereas the first order type does not.

The distributive law is valid, however, if the *second* factor is a sum; i. e.,

$$\gamma)\qquad \mu(\nu + \pi) = \mu\nu + \mu\pi.$$

For, let the order types be represented by the sets $\mathfrak{M}(m)$, $\mathfrak{N}(n)$, $\mathfrak{P}(p)$, where $\mathfrak{N}\cdot\mathfrak{P} = 0$. Then $\mu(\nu + \pi)$ is represented by the set of all elements (q, m), where q is an arbitrary one of the elements n, p, and the relation

$$(q_1, m_1) \prec (q_2, m_2)$$

holds if

$$q_1 \in \mathfrak{N}, \quad q_2 \in \mathfrak{P}$$

or

$$q_1, q_2 \in \mathfrak{N} \text{ and } q_1 \prec q_2 \text{ in } \mathfrak{N}$$

or

$$q_1, q_2 \in \mathfrak{P} \text{ and } q_1 \prec q_2 \text{ in } \mathfrak{P}$$

or

$$q_1 = q_2, \quad m_1 \prec m_2.$$

Now it is evident that the same ordered set is obtained as representative of $\mu\nu + \mu\pi$.

It is customary to denote by λ the order type of the set of all real numbers ordered according to increasing magnitude, and by η, the set of all rational numbers ordered in analogous fashion. Further, the set of real numbers of an open interval, and the set of rational numbers of an open interval, ordered according to magnitude, have the respective order types λ and η too. This is shown by the mapping $y = x/(1 + |x|)$ of the real numbers x on the interval $|y| < 1$. The real numbers of the intervals

$$\langle 0, 1), \ (0, 1\rangle, \ \langle 0, 1\rangle$$

accordingly have the order types

$$1 + \lambda, \quad \lambda + 1, \quad 1 + \lambda + 1,$$

and the rational numbers lying in these intervals have the order types

$$1 + \eta, \quad \eta + 1, \quad 1 + \eta + 1.$$

If we represent $\lambda + 1$ by the interval $(0, 1\rangle$, and a second term λ by the interval $(1, 2)$, we obtain

$$\lambda + 1 + \lambda = \lambda.$$

By a similar juxtaposition of the intervals which represent the terms, it follows that

$$1 + \lambda = (1 + \lambda)n = (1 + \lambda)\omega,$$

$$\lambda + 1 = (\lambda + 1)n = (\lambda + 1)\cdot {}^*\omega,$$

$$\lambda = (\lambda + 1)\omega = (1 + \lambda)\cdot {}^*\omega;$$

and, analogously,

$$\eta + 1 + \eta = \eta$$

$$1 + \eta = (1 + \eta)n = (1 + \eta)\omega,$$

$$\eta + 1 = (\eta + 1)n = (\eta + 1)\cdot {}^*\omega,$$

$$\eta = (\eta + 1)\omega = (1 + \eta)\cdot {}^*\omega.$$

For the order type η, we have, moreover:

$$\eta = \eta n = \eta\omega = \eta \cdot {}^*\omega.$$

This is seen immediately by putting, e. g., intervals with irrational end points in a row.

It follows from the validity of the associative law, that we can speak of a product of a finite number of order types $\mu_1 \cdot \mu_2 \cdots \mu_n$ without inserting parentheses, because, according to the associative law, it is immaterial where they stand. The proof of the associative law shows at the same time, that Definition 2 of the product can be extended so as to provide a general definition for a product of n order types. If the same order type is chosen for all the factors of the product, we get the power μ^n of an order type with a *finite* exponent. From α) then follows:

δ) For finite n, $|\mu^n| = |\mu|^n$.

The power ω^2 can be represented, e. g., by the ordered set

$$\{1, 2, 3, \cdots; \tfrac{1}{2}, \tfrac{3}{2}, \tfrac{5}{2}, \cdots; \tfrac{1}{3}, \tfrac{2}{3}, \tfrac{4}{3}, \cdots; \cdots\}$$

of positive rational numbers.

For the extension of the product to an infinite set of order types, consult Hausdorff [1], p. 147 or [2], p. 73. This extension requires the theory of well-ordered sets.

5. *Power of Type Classes*

It does not yet follow from the results derived thus far, that to every cardinal number $\mathfrak{m}$ there corresponds at least one order type μ with $|\mu| = \mathfrak{m}$, or, in other words, that every set can be ordered. Consequently, if we denote by $\mathfrak{T}_\mathfrak{m}$ the set of distinct order types which belong to a given cardinal number $\mathfrak{m}$, and call this set the type class belonging to $\mathfrak{m}$, we must at first reckon with the possibility that $\mathfrak{T}_\mathfrak{m} = 0$. From above, however, we can already obtain an estimate for the power of the type class. For there is the following

THEOREM 1. *For the type class $\mathfrak{T}_\mathfrak{m}$ belonging to the cardinal number $\mathfrak{m}$ we have* $|\mathfrak{T}_\mathfrak{m}| \leq 2^{\mathfrak{m}\cdot\mathfrak{m}}$. [1]

---

[1] Actually $|\mathfrak{T}_\mathfrak{m}| = 2^\mathfrak{m}$; see Hausdorff [1], p. 455.

Proof: Let $\mathfrak{M}$ be an arbitrary, but then fixed, set with cardinal number $\mathfrak{m}$. Let μ be any order type of $\mathfrak{T}_{\mathfrak{m}}$. Then it is always possible to order $\mathfrak{M}$ so that $\mathfrak{M}$ acquires the order type μ (Theorem 1 on p. 56). The set of all the different order types of the class $\mathfrak{T}_{\mathfrak{m}}$ is therefore a subset of the set of all possible orderings of our set $\mathfrak{M}$. If, now, we have any ordering of $\mathfrak{M}$, we can form the set of all pairs of elements $(m, \overline{m})$ of $\mathfrak{M}$ for which the relation $m \prec \overline{m}$ holds under the ordering which is being considered. The set of *these* pairs of elements is a subset of the set of *all* pairs of elements $(m, \overline{m})$ which can be formed from the elements of the set $\mathfrak{M}$. The set of all possible orderings of the set $\mathfrak{M}$, and *a fortiori* the type class $\mathfrak{T}_{\mathfrak{m}}$, therefore has at most the power which belongs to the set of all subsets which can be formed from the set of all pairs of elements of $\mathfrak{M}$. Now the set of all pairs of elements of $\mathfrak{M}$ has (see the definition on p. 42) cardinal number $\mathfrak{m}^2$, and consequently the set of all subsets of this set has cardinal number $2^{\mathfrak{m} \cdot \mathfrak{m}}$, which proves the assertion.

It follows, in particular, from the theorem, for $\mathfrak{m} = \mathfrak{a}$, that

$$| \mathfrak{T}^{\mathfrak{a}} | \leq 2^{\mathfrak{a} \cdot \mathfrak{a}} = 2^{\mathfrak{a}} = \mathfrak{c}.$$

For this case, however, it is easy to show that the equality sign must actually hold:

THEOREM 2. *For the class* $\mathfrak{T}_{\mathfrak{a}}$ *of order types of enumerable sets* $| \mathfrak{T}_{\mathfrak{a}} | = \mathfrak{c}$.

Proof: There remains to be shown merely that $| \mathfrak{T}_{\mathfrak{a}} | \geq \mathfrak{c}$. For this purpose we make use of the order type $\nu = {}^*\omega + \omega$, which has neither a first nor a last element. With any sequence of natural numbers

$$(1) \qquad (a_1 , a_2 , a_3 , \cdots)$$

we form, by addition, the order type

$$(2) \qquad \alpha = a_1 + \nu + a_2 + \nu + a_3 + \nu + \cdots .$$

We shall show that distinct sequences (1) always give rise to distinct order types (2). Let

$$(3) \qquad (b_1, b_2, b_3, \cdots)$$

also be a sequence of natural numbers, and with it form the order type

$$\beta = b_1 + \nu + b_2 + \nu + b_3 + \nu + \cdots .$$

Suppose that $\alpha = \beta$. Then we are to show that the sequences (1) and (3) coincide. This will follow by induction from the following auxiliary consideration:

Let $\mathfrak{E}_1$, $\mathfrak{E}_2$ and $\mathfrak{N}_1$, $\mathfrak{N}_2$ be ordered sets. Then the relations

$$\mathfrak{E}_1 \simeq \mathfrak{E}_2, \quad \mathfrak{N}_1 \simeq \mathfrak{N}_2$$

follow from

$$\text{(a)} \qquad \begin{cases} \mathfrak{E}_1 + \mathfrak{N}_1 \simeq \mathfrak{E}_2 + \mathfrak{N}_2, \\ \mathfrak{N}_1 \text{ and } \mathfrak{N}_2 \text{ have no first element,} \\ \mathfrak{E}_1 \text{ and } \mathfrak{E}_2 \text{ are finite sets,} \end{cases}$$

as well as from

$$\text{(b)} \qquad \begin{cases} \mathfrak{N}_1 + \mathfrak{E}_1 \simeq \mathfrak{N}_2 + \mathfrak{E}_2, \\ {}_{|}\mathfrak{N}_1{}^{|} = {}_{|}\mathfrak{N}_2{}^{|} = \nu. \end{cases}$$

For consider case (a). Under a similarity mapping of $\mathfrak{E}_1 + \mathfrak{N}_1$ on $\mathfrak{E}_2 + \mathfrak{N}_2$, no element of $\mathfrak{N}_2$ can correspond to an element of $\mathfrak{E}_1$, because every element of $\mathfrak{E}_1$ is preceded by only a finite number of elements in $\mathfrak{E}_1 + \mathfrak{N}_1$, whereas every element of $\mathfrak{N}_2$ is preceded by infinitely many elements in $\mathfrak{E}_2 + \mathfrak{N}_2$. It follows, likewise, that no element of $\mathfrak{N}_1$ can correspond to an element of $\mathfrak{E}_2$. Consequently $\mathfrak{E}_1$ is mapped on $\mathfrak{E}_2$ and $\mathfrak{N}_1$ is mapped on $\mathfrak{N}_2$. In case (b) too, under a similarity mapping of $\mathfrak{N}_1 + \mathfrak{E}_1$ on $\mathfrak{N}_2 + \mathfrak{E}_2$, an element e_1 of $\mathfrak{E}_1$ cannot correspond to an element n_2 of $\mathfrak{N}_2$; otherwise the subset $\mathfrak{N}_1$ preceding e_1 in $\mathfrak{N}_1 + \mathfrak{E}_1$ would have to be mapped on a part $\overline{\mathfrak{N}}_2$ of the elements of $\mathfrak{N}_2$, which is impossible, because $\mathfrak{N}_1$ has no last element, whereas $\overline{\mathfrak{N}}_2$ does. In like manner we see that no element of N_1 can correspond to an element of E_2. Hence, also in this case, E_1 is mapped on E_2 and N_1 is mapped on N_2, Q.E.D.

If we now apply (a) to the equal sums α and β, we get

$$a_1 = b_1$$

and

$$\nu + a_2 + \nu + a_3 + \cdots = \nu + b_2 + \nu + b_3 + \cdots,$$

so that, by (b),

$$a_2 + \nu + a_3 + \cdots = b_2 + \nu + b_3 + \cdots,$$

and hence, again by (a),

$$a_2 = b_2 \quad \text{and} \quad \nu + a_3 + \nu + \cdots = \nu + b_3 + \nu + \cdots,$$

etc. Thus the two sequences (1) and (3) do in fact coincide.

Since the totality of distinct sequences (1) gives rise exclusively, as we have seen, to distinct order types (2), and since (2) implies that

$$|\alpha| \leq \mathfrak{a} \cdot \mathfrak{a} = \mathfrak{a},$$

$\mathfrak{T}_\mathfrak{a}$ has at least the power of the set of all sequences (1) that can be formed from the natural numbers. The power of these sequences of numbers is $\mathfrak{a}^\mathfrak{a} = \mathfrak{c}$, and this proves the theorem.

6. *Dense Sets*

In this paragraph and the next, we shall take up several classifications of order types according to their structure. It is an interesting fact that many concepts formed at first only for point sets can be carried over in general to ordered sets. This will give us means for proving in many cases the dissimilarity of ordered sets. It will also be possible to describe the structure of the set of all rational numbers on a line, and, above all, that of the continuum, solely with the aid of the concept of set.

By the border elements of an ordered set we shall mean its first and last elements, if it has such elements.

An ordered set is called unbordered, if it is not empty and has no border elements.

An unbordered set is always infinite, because in every finite set there is a first as well as a last element.

THEOREM 1. *An unbordered set can be similar only to an unbordered set. Formulated more briefly (cf. Definition 2 on p. 57): if $\mathfrak{M}$ is unbordered, then so is ${}_{|}\mathfrak{M}^{|}$.*

This is merely another formulation of Theorem 2 on p. 56.

Examples of unbordered sets are: The interval (0, 1); the entire number axis; the rational numbers in the interval (0, 1); the type $^*\omega + \omega$.

If two elements m_1 and m_2 of an ordered set $\mathfrak{M}$ have the property that no element of $\mathfrak{M}$ lies between them, then they are called neighboring (consecutive) elements. If these elements satisfy the relation $m_1 \prec m_2$, then m_1 is called the immediate predecessor of m_2, and m_2 is called the immediate successor of m_1. An ordered set is said to be dense, if it contains at least two elements and no neighboring elements.

A dense set is always infinite, because every finite set containing at least two elements has also neighboring elements.

THEOREM 2. *A dense set can be similar only to a dense set; i. e., if $\mathfrak{M}$ is dense, then so is ${}_{|}\mathfrak{M}^{|}$.*

Proof: Let $\mathfrak{M} \simeq \mathfrak{N}$, and suppose that $\mathfrak{N}$ contains two neighboring elements n_1, n_2. Then there can be no element of $\mathfrak{M}$, either, between the elements m_1 and m_2 of $\mathfrak{M}$ which correspond to n_1 and n_2 under a similarity mapping of $\mathfrak{N}$ on $\mathfrak{M}$, since otherwise there would also have to be an element of $\mathfrak{N}$ between n_1 and n_2.

Examples. Dense sets are: The numbers of the interval (0, 1), ordered according to magnitude; the rational numbers in the interval $\langle 0, 1\rangle$, in their natural order; the set (0, 1) + 2. Sets which are not dense are: $\{1, 2, 3, \cdots\}$; the rational numbers when arranged in a sequence, since to every element there now corresponds an immediate successor; the set $\langle 0, 1\rangle + 2$, because the elements 1 and 2 are neighboring.

If we confine ourselves to *enumerable* dense sets, there is the following important

Theorem 3. *All unbordered, dense, enumerable sets are similar to one another*. (G. Cantor)

Proof: Let $\mathfrak{M}(m)$ and $\mathfrak{N}(n)$ be two sets of the kind mentioned in the theorem. Since they are enumerable, they can be written as sequences, whereby their given order may, of course, be disturbed. When written as sequences, let them be

$$\overline{\mathfrak{M}} = \{m', m'', m''', \cdots\} \qquad \text{and} \qquad \overline{\mathfrak{N}} = \{n', n'', n''', \cdots\}.$$

We have to show that the sets $\mathfrak{M}$ and $\mathfrak{N}$ with their original orders can be mapped on each other. To this end, let $m_1 = m'$ correspond to any element n_1 of $\mathfrak{N}$ (say $n_1 = n'$).

We then look for the element in $\overline{\mathfrak{N}} - \{n_1\}$ with smallest upper index, and call this element n_2. In $\mathfrak{N}$, now, either $n_1 \prec n_2$ or $n_1 \succ n_2$. According as the first or second is the case, we choose in $\mathfrak{M}$ an element m_2 with $m_1 \prec m_2$ or $m_1 \succ m_2$; this is possible because $\mathfrak{M}$ is unbordered. We associate this m_2 with n_2.

Now we go over once more to the set $\mathfrak{M}$, and in $\overline{\mathfrak{M}} - \{m_1, m_2\}$ we choose the element with smallest upper index, calling this element m_3. It may appear before, after, or between the elements m_1 and m_2 under the given order of $\mathfrak{M}$. In each case, since $\mathfrak{N}$ is unbordered and dense, there is an element n_3 which has the same position relative to n_1 and n_2 as m_3 has relative to the elements m_1 and m_2.

Now we consider $\mathfrak{N}$ again, choose in $\overline{\mathfrak{N}} - \{n_1, n_2, n_3\}$ the element with smallest upper index, call it n_4, and select an element m_4 of $\mathfrak{M}$ which has the same position relative to the three elements m_1, m_2, m_3 as n_4 has relative to the three elements n_1, n_2, n_3.

In this way, every element of $\overline{\mathfrak{M}}$, i. e., of $\mathfrak{M}$, and every element of $\overline{\mathfrak{N}}$, i. e., of $\mathfrak{N}$, is chosen. Thus, the set $\mathfrak{M}$ is mapped on the set $\mathfrak{N}$, without disturbing the order of succession of the elements. For let $m_1 \prec m_2$ be two arbitrary elements of $\mathfrak{M}$; and let n_1 and n_2 be the corresponding elements of $\mathfrak{N}$. Then the correspondence between one of the pairs of elements m_1, n_1 and m_2, n_2 is set up, according to our procedure, later than that between the other pair; let us assume that it takes place later for the second pair than for the first. Now in making this correspondence between m_2 and n_2 we make sure that the elements m_1 , m_2 have the same relative position in $\mathfrak{M}$ as the elements n_1 , n_2 have in $\mathfrak{N}$. Hence we have actually $\mathfrak{M} \simeq \mathfrak{N}$.

By means of Theorem 3, the order type η of the set of all rational numbers in their natural order is described, and this without reference to these numbers themselves. For, according to Theorem 3, the order type η belongs to every ordered set $\mathfrak{M}$ possessing the following properties:

α) $\mathfrak{M}$ is enumerable;

β) $\mathfrak{M}$ contains neither a first nor a last element;

γ) $\mathfrak{M}$ is dense, i. e., between every pair of its elements there is at least one further element of $\mathfrak{M}$.

Theorem 3 can be generalized to

THEOREM 4. *The four types*

$$\eta, \quad 1 + \eta, \quad \eta + 1, \quad 1 + \eta + 1$$

are the only distinct order types of dense, enumerable sets. That is, every dense, enumerable set is similar to the naturally ordered set of rational numbers in one of the intervals $(0, 1)$, $\langle 0, 1)$, $(0, 1\rangle$, $\langle 0, 1\rangle$.

Proof: By deleting its border elements, every enumerable, dense set can be made into one of the sets treated in Theorem 3, i. e., into a set of type η. If we now replace the border elements, the assertion follows.

The method of proof of Theorem 3 also yields

THEOREM 5. *Given any dense set and any ordered enumerable set, there is always a subset of the former which is similar to the latter.*

Proof: Let the given dense set be made into an unbordered dense set $\mathfrak{M}$ by leaving out any border elements, and let $\mathfrak{A}$ be the given ordered, enumerable set. We have to show that $\mathfrak{M}$ contains a subset which, under the order prescribed for it by $\mathfrak{M}$, is similar to $\mathfrak{A}$. Let us write the set $\mathfrak{A}$ again as a sequence

$$\overline{\mathfrak{A}} = \{a_1 , a_2 , a_3 , \cdots\},$$

disturbing, possibly, the given order. With a_1 we associate an arbitrary element m_1 of $\mathfrak{M}$. With a_2 we associate an element m_2 of $\mathfrak{M}$ which has the same position relative to m_1 as a_2 has to a_1 under the order given by $\mathfrak{A}$ for these elements. Then we associate with a_3 an element m_3 which has the same position relative to m_1, m_2 as a_3 has to a_1, a_2; etc. It is seen immediately from the proof of Theorem 3, that all this is possible and leads to our objective, since the present circumstances are actually simpler than before.

The preceding theorems imply the following:

a) The set of all terminating decimal fractions >0 (or terminating dyadic or n-adic fractions) has order type η. This follows from Theorem 3, because this set is enumerable, unbordered, and dense.

b) For every finite or enumerable order type $\kappa \neq 0$, we have $\eta\kappa = \eta$.

For, represent κ by $\mathfrak{K}(k)$ and η by $\mathfrak{R}(r)$. Then $\eta\kappa$ is represented by the lexicographically ordered set of pairs of elements (k, r). Let (k, r_0) be one of these pairs of elements. Since $\mathfrak{R}$ is unbordered, there is an $r_1 \prec r_0$ and an $r_2 \succ r_0$. Then also $(k, r_1) \prec (k, r_0) \prec (k, r_2)$. Suppose, further, that $(k_1, r_1) \prec (k_2, r_2)$. Then there are two possibilities. If $k_1 \prec k_2$, choose an $r_3 \prec r_2$ and form (k_2, r_3). If, however, $k_1 = k_2$ and $r_1 \prec r_2$, then, since $\mathfrak{R}$ is dense, there is an $r_1 \prec r_3 \prec r_2$, and we again form (k_2, r_3). In each case the newly formed pair of elements lies between the given ones. The set of pairs of elements (k, m) is thus unbordered and dense and, finally, also enumerable, because $|\eta\kappa| = |\eta||\kappa| \leq \mathfrak{a}^2 = \mathfrak{a}$. This set therefore has order type η.

This generalizes a part of the results on p. 66; in particular, it follows that $\eta \cdot \eta = \eta$, and hence also $\eta^3 = \eta$; etc.

c) Likewise: For every finite or enumerable order type $\kappa \neq 0$, we have $(1 + \eta)\kappa = 1 + \eta$ or η according as κ has a first element or not. Thus, in particular, $(1 + \eta)\eta = \eta$, $(1 + \eta)^2 = 1 + \eta$.

7. *Continuous Sets*

Let there be given an ordered set $\mathfrak{M}$. A subset $\mathfrak{A}$ of $\mathfrak{M}$ is called an initial part of $\mathfrak{M}$, if $a \in \mathfrak{A}$ implies that every element preceding a in the set $\mathfrak{M}$ also belongs to $\mathfrak{A}$. The initial part $\mathfrak{A}$ shall always be ordered in the order prescribed by $\mathfrak{M}$. A subset $\mathfrak{Z}$ of $\mathfrak{M}$ is called a remainder of $\mathfrak{M}$, if $z \in \mathfrak{Z}$ implies that every element succeeding z in the set $\mathfrak{M}$ also belongs to $\mathfrak{Z}$. The remainder too shall always be ordered in the order prescribed by $\mathfrak{M}$.

If, e. g., $\mathfrak{R}(r)$ is the set of all rational numbers in the natural order, then the set of rational numbers r for which $r < 2$, constitutes an initial part, and so does the set of rational numbers for which $r \leq 2$. Likewise, a remainder of $\mathfrak{R}$ is given by the set of all $r > 2$ as well as by the set of $r \geq 2$. The empty set is both an initial part and a remainder of every ordered set.

A decomposition of an ordered set $\mathfrak{M}$ is a representation of it in the form $\mathfrak{M} = \mathfrak{A} + \mathfrak{Z}$, where $\mathfrak{A}$ is an initial part and $\mathfrak{Z}$ is a remainder of $\mathfrak{M}$, $\mathfrak{A} \cdot \mathfrak{Z} = 0$, and the sum is, of course, an ordered sum.

A decomposition $\mathfrak{M} = \mathfrak{A} + \mathfrak{Z}$ of an ordered set $\mathfrak{M}$ into two nonempty sets $\mathfrak{A}$ and $\mathfrak{Z}$ is called

a jump, if $\mathfrak{A}$ has a last and $\mathfrak{Z}$ has a first element;
a cut, if $\mathfrak{A}$ has a last and $\mathfrak{Z}$ has no first element, or
if $\mathfrak{A}$ has no last and $\mathfrak{Z}$ has a first element;
a gap, if $\mathfrak{A}$ has no last and $\mathfrak{Z}$ has no first element.

For example, if the real numbers in the following intervals are in natural order,

$$\langle 0, 1\rangle + \langle 2, 3\rangle \text{ is a jump,}$$

$$\langle 0, 1) + \langle 1, 2) \text{ is a cut,}$$

$$(0, 1) + (1, 2) \text{ is a gap.}$$

In the domain of the naturally ordered rational numbers r, with notation that is self-explanatory,

$$\{r \leq 0\} + \{r > 0\} \text{ is a cut,}$$

whereas

$$\{r < \sqrt{2}\} + \{r > \sqrt{2}\} \text{ is a gap.}$$

The existence of jumps in an ordered set is synonymous with the existence of neighboring elements. A dense set, consequently, is a set without jumps; it has only cuts or gaps.

THEOREM 1. *If one of two mutually similar sets has jumps, or cuts, or gaps, then the same holds also for the other set. Hence, given any one of these three properties,* all *sets having the same order type either do or do not possess this property.*

Proof: This follows from the fact that, under a similarity mapping, since it preserves the order of succession of the elements, every decomposition goes over into a decomposition, and a first or last element of a remainder or an initial part, respectively, is mapped again into such an element.

From this it now follows, e. g., that the set of all points $P(x, y)$ in the plane, ordered in g) on p. 57, is not similar to the naturally ordered set $(0, 1)$. For, the first set has gaps, one of which is obtained, e. g., by taking $\mathfrak{A}$ to be the set of points $P(x, y)$ for which $x < 0$, and denoting by $\mathfrak{Z}$ the remainder, determined by $\mathfrak{A}$, of the set. Since the set $(0, 1)$ has no gaps, this set is not similar to the first.

An ordered set, each of whose decompositions is a cut, is said to be continuous. In other words, a set is continuous, if it contains at least two elements, and has neither jumps nor gaps.

Accordingly, a dense set is continuous, if it has no gaps. The set of rational numbers is not continuous, no matter how it is ordered. For, every continuous set is dense, and, by p. 72, there are only four types, η, $1 + \eta$, $\eta + 1$, $1 + \eta + 1$, of dense enumerable sets, each of which has gaps, as was shown in an example above. On the other hand, the set of real numbers in their natural order is continuous.

The concepts which we have introduced here go back essentially to Dedekind. With their aid, the irrational numbers can, as we shall now sketch briefly, be introduced as follows: Suppose that up to now only the rational numbers have been known. With every rational number we can associate a definite point of a given line by means of one of the well-known elementary geometrical constructions. But it is not possible, conversely, to associate a rational number with every point of this line. For if we construct (Fig. 6) an isosceles right triangle

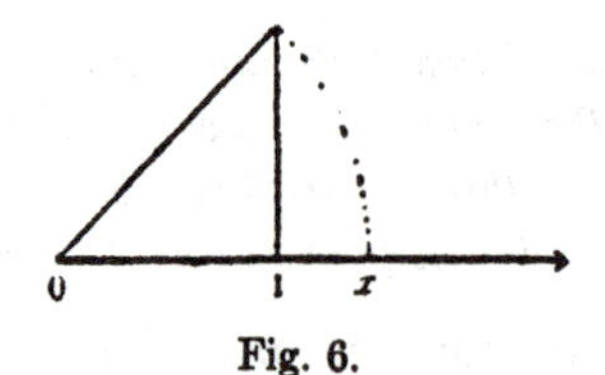

Fig. 6.

with legs of unit length, and mark off the length of the hypotenuse on the given line, beginning at the origin, we obtain a point whose distance x from the origin must satisfy the equation $x^2 = 2$. Such a number x, however, does not exist in the domain of rational numbers. But if one wishes to be able to pursue the study of analytic geometry, it is necessary not only that a point correspond to every number, but also that a number correspond to every point on the line. It is therefore necessary to fill in the "gaps" in the domain of rational numbers (jumps, of course, do not appear in the set of rational numbers). And this takes place according to Dedekind, by introducing the "gaps" as a new kind of number, for which the rules of operation are then defined in a suitable manner. When this has been done, the set of naturally ordered numbers is a continuous set just as the set of points on a straight line. A mapping of the set of numbers on the set of points, preserving the order of their respective elements, as is required in analytic geometry, is now possible.

If Dedekind's reflections are carried over to general ordered sets, we arrive at the following two theorems:

THEOREM 2. *Every continuous set $\mathfrak{M}$ contains a subset of type λ, i. e., a subset which is similar to the set of real numbers in their natural order.*

Proof: According to p. 72, Theorem 5, $\mathfrak{M}$ contains a subset $\mathfrak{R}$ of type η. The set $\mathfrak{R}$ has gaps. Let

$$\mathfrak{R} = \mathfrak{A} + \mathfrak{Z} \tag{1}$$

be a decomposition of $\mathfrak{N}$ which determines a gap in $\mathfrak{N}$. Then $\mathfrak{M}$ contains an element m which, in $\mathfrak{M}$, succeeds all elements of $\mathfrak{A}$ and, at the same time, precedes all elements of $\mathfrak{Z}$. For let $\overline{\mathfrak{A}}$ be the totality of elements of $\mathfrak{M}$ which belong to $\mathfrak{A}$ or precede an element of $\mathfrak{A}$, and let $\overline{\mathfrak{Z}}$ be the totality of elements of $\mathfrak{M}$ which belong to $\mathfrak{Z}$ or succeed an element of $\mathfrak{Z}$. If $\overline{\mathfrak{A}} + \overline{\mathfrak{Z}}$ were a decomposition of $\mathfrak{M}$, $\mathfrak{M}$ would have a gap, contrary to the assumption that $\mathfrak{M}$ is continuous. Hence, $\overline{\mathfrak{A}} + \overline{\mathfrak{Z}}$ lacks at least one element m of $\mathfrak{M}$, and, according to the construction of $\overline{\mathfrak{A}} + \overline{\mathfrak{Z}}$, this element succeeds $\overline{\mathfrak{A}}$ and precedes $\overline{\mathfrak{Z}}$, so that it also succeeds $\mathfrak{A}$ and precedes $\mathfrak{Z}$.

Every gap (1) thus determines at least one element m of $\mathfrak{M}$. For every gap (1), we choose one of these elements m and keep it fixed. Denote by $\overline{\mathfrak{N}}$ the set $\mathfrak{N}$ after it has been supplemented by the addition of these gap elements m. This set, as a subset of the ordered set $\mathfrak{M}$, is also ordered, and is similar to the set of all rational numbers together with their gap elements in the domain of real numbers. In other words, according to Dedekind's introduction of the real numbers, $\overline{\mathfrak{N}}$ is similar precisely to the set of all real numbers, which proves the theorem.

Just as it was possible to describe, on p. 72, the order type of the set of rational numbers without referring to the rational numbers themselves, the analogue is now possible also for the set of real numbers, as the following theorem shows:

THEOREM 3. *Let $\mathfrak{M}$ be an ordered set with the following properties:*

α) $\mathfrak{M}$ is unbordered;

β) $\mathfrak{M}$ is continuous;

γ) $\mathfrak{M}$ contains an enumerable set $\mathfrak{A}$ such that between every pair of elements of $\mathfrak{M}$ there is at least one element of $\mathfrak{A}$; i. e., there is an enumerable set $\mathfrak{A}$ which is dense in $\mathfrak{M}$.

Then $\mathfrak{M}$ has order type λ; i. e., $\mathfrak{M}$ is similar to the naturally ordered set of all real numbers.

Proof: The set $\mathfrak{A}$ is unbordered, because before and after every element of $\mathfrak{A}$ there are still elements of $\mathfrak{M}$, and hence, due to γ), also elements of $\mathfrak{A}$. Likewise, it follows from β) and

γ) that $\mathfrak{A}$ is dense too. Consequently, $\mathfrak{A}$ has order type η, so that it is similar to the set of all rational numbers. Every gap in the set $\mathfrak{A}$ determines, according to the proof of Theorem 2, at least one element m of $\mathfrak{M}$, and, because of γ), not more than one element of $\mathfrak{M}$. This, however, implies the assertion. For, the set of rational numbers can be mapped on $\mathfrak{A}$, whereby with every gap in the domain of rational numbers there is associated precisely one gap of the set $\mathfrak{A}$, and, consequently, also precisely one element of $\mathfrak{M}$. We have thus obtained a similarity mapping of the set of all real numbers on the set $\mathfrak{M}$.

P. Franklin (Transactions of the American Mathematical Society, vol. 27 (1925), pp. 91–100) has made some interesting investigations in connection with our theorems of the last two paragraphs. Suppose that in each of the intervals $0 < x < 1$ and $0 < y < 1$ we have an enumerable set which is dense in that interval. This means that in every subinterval, however small, of the interval (0, 1), there is at least one point of the enumerable set. It follows from Theorem 3 on p. 71, that both sets can be mapped similarly on each other. It is then further evident from the considerations made in connection with the last two theorems, that there exists a continuous monotonic function $y = f(x)$ which maps the interval $0 < x < 1$ continuously on the interval $0 < y < 1$, and, at the same time, maps one of the given sets precisely on the other. Franklin, now, has shown that this function $f(x)$ can be chosen to be actually an analytic function, and has obtained, besides other interesting results, the following. In the interval (0, 1), let there be given a strictly monotonic and continuous function which maps the interval (0, 1) on the interval (0, 1) again. Then this function can be approximated arbitrarily closely by a function which is analytic in the interval and which assumes rational values precisely for rational x's.

CHAPTER IV

Well-ordered Sets and Their Ordinal Numbers

1. Definition of Well-ordering and of Ordinal Number

Ordered sets generally lack a property which is possessed by every set of natural numbers. In every set of natural numbers, namely, there is a smallest; i. e., in every subset of the set of natural numbers, ordered according to increasing magnitude of its elements, there is a first element. This property of the natural numbers is used in many proofs (cf. especially §7 in this connection), without, perhaps, one's being always clearly conscious of it. It is therefore to be expected that the corresponding property is also of importance for arbitrary ordered sets. Accordingly, we set down

Definition 1. *An ordered set $\mathfrak{M}$ is said to be well-ordered, if $\mathfrak{M}$ itself, as well as every nonempty subset of $\mathfrak{M}$, has a first element under the order prescribed for its elements by $\mathfrak{M}$. The empty set is also regarded as well-ordered.*

Examples. Every finite set is well-ordered, and so is the set $\{1, 2, 3, \cdots\}$. The set $\{\cdots, 3, 2, 1\}$ is not well-ordered, because it has no first element. The set of real numbers in the interval $\langle 0, 1\rangle$, in their natural order, is not well-ordered. The set itself, to be sure, has a first element, but the subset $(0, 1\rangle$ does not. It follows, likewise, that no set of order type η, $1 + \eta$, $\eta + 1$, or $1 + \eta + 1$, is well-ordered. On the other hand, e. g., the set of positive rational numbers in the order

$$\{1, 2, 3, \cdots; \tfrac{1}{2}, \tfrac{3}{2}, \tfrac{5}{2}, \cdots; \tfrac{1}{3}, \tfrac{2}{3}, \tfrac{4}{3}, \cdots; \cdots\} \tag{1}$$

is well-ordered. For let any subset of this set be given. This subset contains fractions with smallest denominator, and among these there is one fraction with smallest numerator. This fraction, then, is the first element of the subset.

THEOREM 1. *In every well-ordered set, every element which is not a last element has an immediate successor.*

Proof: Let m_1 be an element of the well-ordered set $\mathfrak{M}$. The set of elements coming after m_1 in $\mathfrak{M}$ is a subset of $\mathfrak{M}$, and therefore, according to the definition of well-ordered set, if it is not empty it has a first element m_2. Then $m_1 \prec m_2$, and there is no element of $\mathfrak{M}$ between these two elements, according to the definition of m_2.

Thus, the well-ordered sets certainly do not belong to the dense or the continuous sets considered in the two preceding paragraphs.

Further, the following two facts follow immediately from the definition:

THEOREM 2. *Every subset $\mathfrak{N}$ of a well-ordered set $\mathfrak{M}$ is also well-ordered (under the order prescribed for it by $\mathfrak{M}$).*

THEOREM 3. *Every ordered set which is similar to a well-ordered set is itself well-ordered.*

According to Theorem 3, of the sets having any given order type, either *all* or *none* are well-ordered. This fact renders possible the following

Definition 2. *By an ordinal number we mean an order type which is represented by* well-ordered *sets. Small Greek letters will be used to denote ordinal numbers as well as order types.*

In the case of finite sets, cardinal number, order type, and ordinal number coincide. With infinite sets, to a cardinal number there may correspond many order types. Whether there are always ordinal numbers among these, i. e., whether every set can be well-ordered, will not be decided until later (§11).

It follows already from previous considerations of ours, that there are ordinal numbers corresponding to enumerable sets. For according to the examples on the preceding page, the set $\{1, 2, 3, \cdots\}$ and the set (1) are well-ordered. The first set has order type ω, and the second, by p. 66, has order type ω^2. Hence, these two order types are also ordinal numbers. We shall see in the next paragraph how infinitely many ordinal numbers can already be formed from ω alone.

2. *Addition of Arbitrarily Many, and Multiplication of Two, Ordinal Numbers*

On p. 61 we dealt with ordered addition of ordered sets. If the ordered sets appearing there are replaced by well-ordered sets, we arrive at

THEOREM 1. *Let $\mathfrak{K}(\mathfrak{M}_k)$ be a well-ordered set of mutually exclusive well-ordered sets; i. e., let $\mathfrak{K}(k)$ be a well-ordered set, and with every element k of this set, associate a well-ordered set $\mathfrak{M}_k$, where all these $\mathfrak{M}_k$'s are taken to be mutually exclusive. Then we assert that the ordered sum $\mathfrak{S} = \sum_{k \in \mathfrak{K}} \mathfrak{M}_k$ of these sets, formed according to Definition 3 on p. 61, is itself a well-ordered set.*

Proof: Let $\mathfrak{T}$ be an arbitrary nonempty subset of $\mathfrak{S}$. We have to show that $\mathfrak{T}$ has a first element. To this end we consider the elements $k \in \mathfrak{K}$ such that elements of $\mathfrak{M}_k$ actually appear in $\mathfrak{T}$. These k's constitute a subset of the well-ordered set $\mathfrak{K}$, and they therefore have a first element k_0. Further, the elements of $\mathfrak{M}_{k_0}$ which appear in $\mathfrak{T}$ form a subset of the well-ordered set $\mathfrak{M}_{k_0}$, and they therefore also have a first element m_0. This element, however, according to the order of $\mathfrak{S}$ determined in Definition 3 on p. 61, has the property that it precedes all other elements in $\mathfrak{T}$.

No new definition is necessary for the addition of ordinal numbers, since ordinal numbers are merely special order types, so that the definition of sum given in ch. III, §3 holds for them too. By specializing that definition somewhat, we obtain

THEOREM 2. *Let $\mathfrak{K}(\mu_k)$ be a complex of ordinal numbers arising from a well-ordered set $\mathfrak{K}(k)$—briefly, a well-ordered complex of ordinal numbers. Then its ordered sum $\sigma = \sum_{k \in \mathfrak{K}} \mu_k$ is also an ordinal number.*

Proof: According to Definition 4 on p. 61, $\sigma = |\sum_{k \in \mathfrak{K}} \mathfrak{M}_k|$, where the $\mathfrak{M}_k$'s are mutually exclusive sets which represent the μ_k's. Since the μ_k's are now ordinal numbers, the $\mathfrak{M}_k$'s are well-ordered. $\mathfrak{K}$ is also well-ordered, so that, according to Theorem 1, the set sum is well-ordered, and hence it does indeed represent an ordinal number.

What is important here is that well-ordered addition of ordinal numbers invariably leads again to ordinal numbers. In particular, the sum of two ordinal numbers is always an ordinal number.

If the terms of the well-ordered complex of ordinal numbers are all the same, Theorem 2 gives rise to

THEOREM 3. *The product of two, and hence also of finitely many, ordinal numbers is again an ordinal number. In particular, every power of an ordinal number is an ordinal number, provided that the exponent is a finite number.*

From this it now follows that $\omega \cdot n$ and ω^n are ordinal numbers too. Further, Theorem 2 shows that every "polynomial"

$$\omega^{n_1} \cdot a_{n_1} + \omega^{n_2} \cdot a_{n_2} + \cdots + \omega^{n_r} \cdot a_{n_r} ,$$

whose coefficients a_m are finite numbers, is also an ordinal number.

3. Subsets and Similarity Mappings of Well-ordered Sets

Corresponding to the theorem that every transfinite set contains an enumerable subset, we have the following theorem for well-ordered sets:

THEOREM 1. *Every transfinite well-ordered set contains a subset whose ordinal number is ω.*

Proof: $\mathfrak{M}$ contains a first element m_0 ; further, according to p. 80, an immediate successor m_1 ; then an m_2 which immediately succeeds m_1 ; etc. The process does not terminate, because $\mathfrak{M}$ is transfinite, and therefore leads to an enumerable sequence of elements

$$(1) \qquad \mathfrak{A} = \{m_0 , m_1 , m_2 , \cdots\}$$

having ordinal number ω, and the elements in this sequence appear in the same order of succession as they do in M.

For every transfinite well-ordered set $\mathfrak{M}$, there is a similarity mapping of $\mathfrak{M}$ on one of its subsets $\mathfrak{N}$, such that, if m and n are corresponding elements of $\mathfrak{M}$ and $\mathfrak{N}$, and we consider their position in $\mathfrak{M}$, $m \prec n$ infinitely often. For let us form the set (1), as in the preceding proof. This set is an initial part of $\mathfrak{M}$. Let the corresponding (possibly empty) remainder be $\mathfrak{Z}$, so that $\mathfrak{M} = \mathfrak{A} + \mathfrak{Z}$. Now put

$$\overline{\mathfrak{A}} = \{m_1, m_2, \cdots\}, \qquad \mathfrak{N} = \overline{\mathfrak{A}} + \mathfrak{Z}.$$

The sets $\mathfrak{A}$ and $\overline{\mathfrak{A}}$ are mapped similarly on each other by associating each m_n of the first set with the m_{n+1} of the second set. Further, we let every element of $\mathfrak{Z}$ correspond to itself. This gives a similarity mapping of $\mathfrak{M}$ on $\mathfrak{N}$ under which every element m_n of $\mathfrak{M}$ corresponds to that element of $\mathfrak{N}$ which just succeeds m_n in $\mathfrak{M}$.

The reverse, however, — and this is a very important result for the theory of well-ordered sets—cannot occur. For we have

THEOREM 2. *Let there exist a similarity mapping between the well-ordered set $\mathfrak{M}$ and its (proper or improper) subset $\mathfrak{N}$, denoting corresponding elements of $\mathfrak{M}$ and $\mathfrak{N}$ by m and n. If we consider the position of these elements in the set $\mathfrak{M}$, then we have invariably $m = n$ or $m \prec n$, but never $n \prec m$* (Zermelo).

Proof: Assume the theorem to be false. Then the set of pairs of elements m, n for which $n \prec m$, is not empty. The n's with $n \prec m$ constitute a subset of the well-ordered set $\mathfrak{N}$, and they therefore have a first element n_0. For the element m_0 corresponding to n_0 we have $n_0 \prec m_0$. The element n_0, however, is an element of $\mathfrak{M}$ too, and will in this role be denoted also by m_1. To this element m_1 there corresponds an element n_1 of $\mathfrak{N}$. Thus, with the elements m_0, m_1 of $\mathfrak{M}$ there are associated the elements n_0, n_1 of $\mathfrak{N}$. Since $m_1 = n_0 \prec m_0$, and the similarity mapping does not disturb the order of succession, we have also $n_1 \prec n_0$, i. e., $n_1 \prec m_1$. Hence, n_0 is not the first element of $\mathfrak{N}$ to precede the corresponding element of $\mathfrak{M}$: the element $n_1 \prec n_0$ already has this property. Our assumption that the assertion is false has thus led to the desired contradiction.

From this we now obtain

THEOREM 3. *Every well-ordered set $\mathfrak{M}$ can be mapped similarly on itself only by the identity mapping. If $\mathfrak{M}$ is well-ordered, and $\mathfrak{M} \simeq \mathfrak{N}$, then there is only one similarity mapping of $\mathfrak{M}$ on $\mathfrak{N}$.*

Proof: $\mathfrak{M}$ is a subset of itself. Let m and $\overline{m}$ be two corresponding elements under a mapping of $\mathfrak{M}$ on itself. Then, according to Theorem 2, we never have $m \prec \overline{m}$ or $\overline{m} \prec m$, so that invariably $m = \overline{m}$. This now gives the second assertion. For if $\mathfrak{N}$ could be mapped on $\mathfrak{M}$ in two ways, it would be possible, by interposing $\mathfrak{N}$, to map $\mathfrak{M}$ on itself in two ways, which is impossible.

We have already considered a particular kind of subset of an ordered set, namely, the initial part. We shall now alter the terminology a little. Let $\mathfrak{M}$ be a well-ordered set, and let m be one of its elements. Then the subset of elements of $\mathfrak{M}$ which precede the element m will be called the segment $\mathfrak{M}_m$ determined by m. The segment belonging to the first element of $\mathfrak{M}$ shall be the empty set. The complement of $\mathfrak{M}_m$ in $\mathfrak{M}$ is called the remainder determined by m, and will be denoted by $\mathfrak{M} - \mathfrak{M}_m$. The segment determined by m does not contain the element m itself; this element belongs, rather, to the remainder determined by m. From this we immediately get

THEOREM 4. *If $\mathfrak{N}$ is a segment of the well-ordered set $\mathfrak{P}$, and $\mathfrak{M}$ is a segment of the set $\mathfrak{N}$, then $\mathfrak{M}$ is also a segment of $\mathfrak{P}$.*

THEOREM 5. *Let $\mathfrak{M}_m$ and $\mathfrak{M}_n$ be two distinct segments of the same well-ordered set $\mathfrak{M}$. Then one of the two segments is a segment of the other.*

Proof: Let $m \neq n$, so that, say, $m \prec n$. Then, for every $p \in \mathfrak{M}$ for which $p \prec m$, we have *a fortiori* $p \prec n$, which yields the assertion.

THEOREM 6. *Under a similarity mapping of a well-ordered set $\mathfrak{M}$ on a well-ordered set $\mathfrak{N}$, every segment of $\mathfrak{M}$ is transformed into a segment of $\mathfrak{N}$.*

Proof: Let $\mathfrak{M}_m$ be a segment of $\mathfrak{M}$, and let the element n of

$\mathfrak{N}$ correspond to the element m of $\mathfrak{M}$. Then the elements of $\mathfrak{M}$ which precede the element m go over precisely into the elements of $\mathfrak{N}$ which precede the element n.

From Theorem 2 we obtain further:

THEOREM 7. *No well-ordered set* $\mathfrak{M}$ *is similar to any of its segments, or to a segment of a subset. Two distinct segments of the same well-ordered set are never similar to each other.*

Proof: If, for some segment $\overline{\mathfrak{M}}_m$ of the subset $\overline{\mathfrak{M}} \subseteq \mathfrak{M}$, we had $\overline{\mathfrak{M}}_m \simeq \mathfrak{M}$, the element m of the set $\mathfrak{M}$ would correspond to an element of $\mathfrak{M}_m$, i. e., to an element preceding it in $\mathfrak{M}$, which, by Theorem 2, is impossible. If, now, $\mathfrak{M}_m$ and $\mathfrak{M}_n$ are two distinct segments of the same set, then, according to Theorem 5, one of them must be a segment of the other, and consequently the second assertion follows from the first.

4. *The Comparison of Ordinal Numbers*

For order types, and consequently for ordinal numbers, it is clear when they are equal, but the relations "less than" and "greater than" have not yet been defined. This will now take place for ordinal numbers (but not even now for order types).

Let μ, ν be the ordinal numbers of the well-ordered sets $\mathfrak{M}$, $\mathfrak{N}$. Then we say that $\mu < \nu$ and, at the same time, also $\nu > \mu$, if $\mathfrak{M}$ is similar to a segment of $\mathfrak{N}$.

On account of §3, Theorem 6, this definition is independent of the particular representatives of the ordinal numbers μ and ν.

As for rules governing inequalities, we first obtain almost immediately the following two:

α) If $\mu < \nu$ and $\nu < \pi$, then $\mu < \pi$ (transitive law).

For let the three ordinal numbers be represented by the sets $\mathfrak{M}$, $\mathfrak{N}$, $\mathfrak{P}$. Then the assumptions state that $\mathfrak{M}$ is similar to a segment $\mathfrak{N}_n$ of $\mathfrak{N}$, and $\mathfrak{N}$ is similar to a segment $\mathfrak{P}_p$ of $\mathfrak{P}$. According to §3, Theorem 6, the similarity mapping of $\mathfrak{N}$ on $\mathfrak{P}$, also maps $\mathfrak{N}_n$ on a segment of $\mathfrak{P}$, so that $\mathfrak{M}$ too is similar to a segment of $\mathfrak{P}$.

β) For every pair of ordinal numbers μ, ν, *at most one* of the three relations

$$\mu < \nu, \quad \mu = \nu, \quad \mu > \nu$$

is valid.

For if we had simultaneously $\mu < \nu$ and $\mu = \nu$, we should also have $\mu < \mu$, i. e., a well-ordered set would be similar to one of its segments, which is impossible according to §3, Theorem 7. Likewise it follows that $\mu = \nu$ and $\mu > \nu$ cannot hold simultaneously either. Finally, if we had $\mu < \nu$ and at the same time $\mu > \nu$, it would follow from α) that $\mu < \mu$ again, so that this case too cannot occur.

We have now gotten just as far with ordinal numbers as with cardinal numbers. For both kinds of numbers, namely, the question is still open, whether it is also true that invariably at *least* one of the relations β) holds, i. e., whether every two numbers of the same variety are always comparable with one another. For ordinal numbers, this question can now be decided. From this will then follow the decision for cardinal numbers too, but, to be sure, only on the basis of the very deep well-ordering theorem.

Two ordinal numbers which are represented by segments of one and the same set are certainly comparable. For, according to §3, Theorem 5, either both segments are the same, or one of them is a segment of the other. The general proof is executed with the aid of a certain normal representation of a given ordinal number by means of ordinal numbers themselves.

Let $\mu > 0$. Then the set $\mathfrak{W}_\mu$, which will also be used frequently in the future, is understood to be the set of all ordinal numbers which are less than μ. If $\mu = 0$, we put $\mathfrak{W}_0 = 0$.[1]

If $\mu > 0$, $\mathfrak{W}_\mu$ is not empty, because then the ordinal number 0 always belongs to this set. We have, for example,

$$\mathfrak{W}_n = \{0, 1, 2, \cdots, n-1\}; \quad \mathfrak{W}_\omega = \{0, 1, 2, \cdots\}.$$

THEOREM 1. *The elements of the set* $\mathfrak{W}_\mu$ *are comparable with*

[1]The reader should observe that neither in this definition nor in the proofs of the following two theorems do we make use of a comparability theorem. It is just this theorem, rather, which is to be proved by means of the ensuing considerations.

one another. If, as is accordingly possible, we order the ordinal numbers appearing in $\mathfrak{W}_\mu$ *according to increasing magnitude, the set* $\mathfrak{W}_\mu$ *has ordinal number* μ.

Proof: Let $\mathfrak{M}$ be a well-ordered set with ordinal number μ. Let α be any element of $\mathfrak{W}_\mu$, i. e., let $\alpha < \mu$. Then, by the definition of "less than", α is the ordinal number of a segment of $\mathfrak{M}$. All ordinal numbers appearing in $\mathfrak{W}_\mu$ are thus ordinal numbers of segments of the same set, and hence, by what was said on the preceding page, these ordinal numbers are all comparable with one another. It is clear that the ordinal number of an arbitrary segment of $\mathfrak{M}$ is also contained in the set $\mathfrak{W}_\mu$. Now let α be an element of $\mathfrak{W}_\mu$, and let $\mathfrak{M}_m$ be the segment of $\mathfrak{M}$ which belongs to this ordinal number. We associate the ordinal number α with the element m. By what has preceded, this defines a mapping of the set $\mathfrak{W}_\mu$ on the set $\mathfrak{M}$. But this mapping is also a similarity mapping. For if $\alpha_1 < \alpha_2$, then $\mathfrak{M}_{m_1}$ is a segment of $\mathfrak{M}_{m_2}$, so that $m_1 \prec m_2$. Consequently $\mathfrak{W}_\mu \simeq \mathfrak{M}$, and this completes the proof of the theorem.

With this theorem we have obtained a certain normal representation of ordinal numbers. In all operations, now, where a well-ordered set may be replaced by any set similar to it, we can always replace the given set by the set $\mathfrak{W}_\mu$, where μ is the ordinal number of the given set. This is often done in proofs. We first prove in this manner the comparability theorem for ordinal numbers:

THEOREM 2. *Any two ordinal numbers* μ, ν *satisfy at least one, and consequently, according to* β), *invariably precisely one, of the three relations*

$$\mu < \nu, \quad \mu = \nu, \quad \mu > \nu;$$

i. e., two ordinal numbers are always comparable with one another.

Proof: We make use of the sets $\mathfrak{W}_\mu$ and $\mathfrak{W}_\nu$. Let $\mathfrak{D}$ be their intersection, i. e., the set of ordinal numbers which are both $<\mu$ and $<\nu$. $\mathfrak{D}$, as a subset of the well-ordered set $\mathfrak{W}_\mu$, is well-ordered, and therefore has an ordinal number δ. We shall now show that δ is comparable with μ, and that $\delta \leq \mu$. In demonstrating this, we may assume that $\delta \neq \mu$. Then $\mathfrak{D}$ is a proper subset of $\mathfrak{W}_\mu$, and

$$\mathfrak{W}_\mu = \mathfrak{D} + (\mathfrak{W}_\mu - \mathfrak{D}).$$

Here $\mathfrak{D}$ is an initial part of $\mathfrak{W}_\mu$. For let $\alpha \in \mathfrak{D}$ and $\beta \in (\mathfrak{W}_\mu - \mathfrak{D})$. Then α and β, as elements of $\mathfrak{W}_\mu$, are comparable. They are distinct, however, so that either $\alpha < \beta$ or $\beta < \alpha$. If we had $\beta < \alpha$, then, since $\alpha < \mu, \nu$, we should also have $\beta < \mu, \nu$, so that β too would belong to $\mathfrak{D}$, which is not consistent with the definition of β. Consequently we have always $\alpha < \beta$, i. e., $\mathfrak{D}$ is an initial part of $\mathfrak{W}_\mu$. The subset of those elements of $\mathfrak{W}_\mu$ which do not belong to $\mathfrak{D}$ has a first element $\bar{\beta}$, and $\mathfrak{D}$ is the segment of $\mathfrak{W}_\mu$ determined by $\bar{\beta}$. Since this segment, by Theorem 1, has ordinal number $\bar{\beta}$, $\delta = \bar{\beta}$, and hence $\delta < \mu$. This is valid under the assumption that $\delta \neq \mu$. Hence, invariably $\delta \leq \mu$, and analogously we obtain $\delta \leq \nu$. Now we cannot have $\delta < \mu$, $\delta < \nu$ simultaneously, because this would imply that $\delta \in \mathfrak{D}$, so that δ would have to belong to the segment of $\mathfrak{W}_\mu$ determined by $\bar{\beta} = \delta$, which is impossible. Therefore only the following combinations are possible:

$$\delta = \mu,\ \delta = \nu, \quad \text{so that} \quad \mu = \nu;$$

$$\delta = \mu,\ \delta < \nu, \quad \text{so that} \quad \mu < \nu;$$

$$\delta < \mu,\ \delta = \nu, \quad \text{so that} \quad \mu > \nu;$$

i. e., the case that μ and ν are not comparable does not occur.

From this follows immediately the comparability of cardinal numbers of well-ordered sets:

THEOREM 3. *Let* $\mathfrak{m}$ *and* $\mathfrak{n}$ *be two cardinal numbers which can be represented by well-ordered sets. Then precisely one of the three relations*

$$\mathfrak{m} < \mathfrak{n}, \quad \mathfrak{m} = \mathfrak{n}, \quad \mathfrak{m} > \mathfrak{n}$$

holds.

Proof: On account of β) on p. 19, we have only to show that *at least* one of these relations holds. Let $\mathfrak{m}$, $\mathfrak{n}$ be represented by the well-ordered sets $\mathfrak{M}$, $\mathfrak{N}$ having ordinal numbers μ, ν. If $\mu = \nu$, then *a fortiori* $\mathfrak{m} = \mathfrak{n}$. If, however, $\mu < \nu$, say, then $\mathfrak{M}$ is similar to a segment of $\mathfrak{N}$, i. e., $\mathfrak{M}$ is equivalent to a subset of $\mathfrak{N}$. Hence, by the Corollary on p. 25, $\mathfrak{m} \leq \mathfrak{n}$.

According to this, the comparability of two arbitrary cardinal numbers is ensured, if we can show that every set can be well-ordered. The general comparability will, as a matter of fact, be proved in this way, and this is also the only method up to now by which comparability in general has been demonstrated. For the time being, however, we shall deal further with ordinal numbers, and we first derive two simple facts concerning inequalities.

THEOREM 4. *Let the set $\mathfrak{M}$ be a subset of the well-ordered set $\mathfrak{N}$. Then their ordinal numbers satisfy the relation $\mu \leq \nu$.*

Proof: If the assertion were false, the comparability theorem would yield $\mu > \nu$, so that $\mathfrak{N}$ would be similar to a subset, which is impossible by §3, Theorem 7.

THEOREM 5. *Two ordinal numbers satisfy the relation $\mu < \nu$ if, and only if, there is an ordinal number $\zeta > 0$ such that $\mu + \zeta = \nu$.*

Proof: Let μ and ν be represented by the sets $\mathfrak{M}$ and $\mathfrak{N}$. If $\mu < \nu$, then $\mathfrak{M}$ is similar to a segment $\mathfrak{A}$ of $\mathfrak{N}$. For the remainder $\mathfrak{Z}$ of $\mathfrak{N}$ we have $\mathfrak{A} + \mathfrak{Z} = \mathfrak{N}$. Hence, if $\mathfrak{Z}$ has ordinal number ζ, $\mu + \zeta = \nu$. Here $\zeta > 0$ because $\mathfrak{Z}$ is not empty. Conversely, suppose that $\mu + \zeta = \nu$ and that ζ is represented by the set $\mathfrak{Z}$ which has no elements in common with $\mathfrak{M}$. Then $\mathfrak{M} + \mathfrak{Z}$ has ordinal number $\nu = \mu + \zeta$, and $\mu < \nu$ because $\mathfrak{M}$ is a segment of $\mathfrak{M} + \mathfrak{Z}$.

5. *Sequences of Ordinal Numbers*

Let there be given a well-ordered complex of ordinal numbers; i. e., to every element k of a well-ordered set $\mathfrak{K}(k)$, let there correspond an ordinal number μ_k. Such a complex of ordinal numbers will also be called a sequence of ordinal numbers. In particular, a sequence of ordinal numbers will be called an increasing sequence, if $k_1 \prec k_2$ always implies

$\mu_{k_1} < \mu_{k_2}$; a decreasing sequence, if $k_1 \prec k_2$ invariably implies $\mu_{k_1} > \mu_{k_2}$. Concerning the former we have

THEOREM 1. *Every set of ordinal numbers, if ordered according to increasing magnitude, is well-ordered, and can therefore be written as an increasing sequence.*

Proof: We have merely to show that every nonempty subset $\mathfrak{W}$ of the given set of ordinal numbers has a smallest element. Let μ be an arbitrary ordinal number in $\mathfrak{W}$. If μ is not already the smallest ordinal number in $\mathfrak{W}$, form the well-ordered set $\mathfrak{W}_\mu$. The intersection $\mathfrak{W} \cdot \mathfrak{W}_\mu$ can then be well-ordered, and is not empty, so that it therefore contains a smallest ordinal number, which is obviously also the smallest ordinal number in $\mathfrak{W}$.

Concerning decreasing sequences of ordinal numbers, we have

THEOREM 2. *Every decreasing sequence of ordinal numbers contains only a finite number of ordinal numbers.*

Proof: Suppose that the sequence were not finite. Then, by Theorem 1 on p. 82, the set $\mathfrak{K}(k)$ defining the sequence would contain a subset $\{k_1, k_2, k_3, \cdots\}$ of ordinal number ω. The ordinal numbers $\mu_1, \mu_2, \mu_3, \cdots$ associated with these elements would have no smallest, because they decrease, contradicting Theorem 1.

THEOREM 3. *To every set $\mathfrak{W}$ of ordinal numbers μ, there is an ordinal number which is greater than every ordinal number μ in $\mathfrak{W}$; and, in fact, there is a definite next larger ordinal number ν.*

Proof: We may assume that $\mathfrak{W}(\underline{\mu})$ is an increasing sequence. From $\mathfrak{W}(\mu)$ we form a new set $\mathfrak{W}$ of ordinal numbers by replacing every element μ of $\mathfrak{W}$ by $\mu \underline{+} 1$. Let σ be the sum of the elements of the well-ordered set $\mathfrak{W} = \mathfrak{W}(\mu + 1)$. Then

$$\text{every } \mu + 1 \leq \sigma,$$

as is seen from §4, Theorem 4 by passing from the ordinal numbers to their representatives. Hence, by §4, Theorem 5,

$$\text{every } \mu < \sigma,$$

which proves the first part of the assertion.

If, now, σ is not yet the next larger ordinal number relative to the set $\mathfrak{W}$, i. e., if there are still ordinal numbers which are larger than every μ and smaller than σ, then all such ordinal numbers are contained in the set $\mathfrak{W}_\sigma$. Since this set is well-ordered, the subset of $\mathfrak{W}_\sigma$ consisting of those numbers which are greater than every μ, has a smallest element ν, Q.E.D.

This theorem now yields a result which is similar to the one established on p. 36 for cardinal numbers. To every set of ordinal numbers, namely, there is, according to our theorem, a still greater ordinal number. Accordingly, the "set of *all* ordinal numbers" is a meaningless concept. We shall return to this so-called Burali-Forti paradox too in our concluding remarks. In the meantime, we must note that it is not permissible to operate with the set of *all* ordinal numbers. And the proofs given thus far have been carried out with this in mind. Indeed, it is for this reason that we were always careful to operate with the set $\mathfrak{W}_\mu$, i. e., with the set of all ordinal numbers less than a given ordinal number, instead of with the set of *all* ordinal numbers, which might at first, perhaps, seem more natural.

We shall also derive the following theorem at this time, although it will not be used until later:

THEOREM 4. *Let $\mathfrak{W}$ be an enumerable set of ordinal numbers of enumerable sets. Then the ordinal number which immediately succeeds $\mathfrak{W}$ is also an ordinal number of an enumerable set.*

Proof: Let the elements of $\mathfrak{W}$ again be μ. Then

$$|\sigma| = \sum_{\mu \in \mathfrak{W}} |\mu| \leq \mathfrak{a} \cdot \mathfrak{a} = \mathfrak{a}.$$

Consequently, for the ordinal number which immediately succeeds $\mathfrak{W}$ we also have $|\nu| \leq \mathfrak{a}$.

THEOREM 5. *Let $\mathfrak{W}$ be a set of ordinal numbers μ, and let $\overline{\mathfrak{W}}$ be a set of ordinal numbers $\bar{\mu}$ such that to every μ there is a $\bar{\mu} \geq \mu$.*

Let ν, $\bar{\nu}$ be the ordinal numbers immediately succeeding μ, $\bar{\mu}$, respectively. Then $\bar{\nu} \geq \nu$.

Proof: $\bar{\nu}$ is greater than every $\bar{\mu}$, and is therefore also greater than every μ. ν, however, is the smallest number which is greater than every μ. Hence, $\bar{\nu} \geq \nu$.

If ν is an arbitrary ordinal number, the following two cases are conceivable:

a) To ν there corresponds a next smaller ordinal number, i. e., a $\mu < \nu$ such that no ordinal number lies between μ and ν. By §4, Theorem 5, there exists a number ζ such that $\mu + \zeta = \nu$, and ζ then must necessarily be the smallest positive ordinal number, which is 1; i. e., $\mu + 1 = \nu$. In this case we write also $\mu = \nu - 1$, and ν is called an isolated number or an ordinal number of the first kind.

b) To ν there corresponds no next smaller ordinal number. In this case we write also $\nu = \lim_{\mu<\nu} \mu$, and ν is called a limit number or an ordinal number of the second kind. The limit numbers represent precisely those well-ordered sets without a last element, as consideration of the set $\mathfrak{W}_\nu$ shows.

By a fundamental sequence we mean an increasing sequence of ordinal numbers containing no greatest. Theorem 3 can be applied to every fundamental sequence. This means that to the numbers of a fundamental sequence there corresponds a next larger ordinal number, and this number, according to the definition of fundamental sequence, is certainly not equal to any of the numbers of the sequence increased by 1. But it is still conceivable at first, that the concept "next larger number after the numbers of a fundamental sequence" does not completely coincide with the concept of limit number. That they actually do coincide, however, is shown by

THEOREM 6. *An increasing sequence of ordinal numbers determines a limit number as the next larger number in the sense of Theorem 3, if, and only if, it is a fundamental sequence. This next larger number of the fundamental sequence $\mathfrak{W}$ is then called the limit of the fundamental sequence, and is denoted by $\nu = \lim_{\mu \in \mathfrak{W}} \mu$.*

Proof: If the number ν which, by virtue of Theorem 3, immediately succeeds the increasing sequence $\mathfrak{W}(\mu)$, is a limit number, then $\mathfrak{W}$ is a fundamental sequence. For if $\mathfrak{W}$ contained a greatest number $\bar{\mu}$, there would be additional numbers between the limit number ν and $\bar{\mu}$, contrary to the definition of ν. On the other hand, let ν be determined by a fundamental sequence $\mathfrak{W}(\mu)$. We are to show that, for every number $\rho < \nu$, there are additional ordinal numbers between ρ and ν. Now there is actually always a μ in $\mathfrak{W}$ such that $\mu > \rho$; for otherwise we should have invariably $\mu \leq \rho$, and hence, since there is no greatest number among the μ's, invariably $\mu < \rho$, and consequently $\nu \leq \rho$.

6. *Operating With Ordinal Numbers*

Since we have a comparability theorem for ordinal numbers, we can set up much more far-reaching rules of operation for them than for order types. We shall now derive these rules.

a) If $\mu < \nu$, and ρ is an arbitrary ordinal number, then

α) $\rho + \mu < \rho + \nu$;

β) $\mu + \rho \leq \nu + \rho$;

γ) $\rho \cdot \mu < \rho \cdot \nu$ for $\rho > 0$;

δ) $\mu \cdot \rho \leq \nu \cdot \rho$.

Proof: From the assumption follows the existence of a ζ such that $\mu + \zeta = \nu$. Consequently,

α) $\rho + \nu = \rho + (\mu + \zeta) = (\rho + \mu) + \zeta > \rho + \mu$;

γ) $\rho \cdot \nu = \rho(\mu + \zeta) = \rho\mu + \rho\zeta > \rho\mu$.

To prove the two remaining rules, represent the ordinal numbers μ, ν, ρ by the sets $\mathfrak{M}$, $\mathfrak{N}$, $\mathfrak{R}$. Since $\mu < \nu$, we can assume that $\mathfrak{M} \subset \mathfrak{N}$. Further, let $\mathfrak{N} \cdot \mathfrak{R} = 0$. Then, for the ordered sums and products,

$$\mathfrak{M} + \mathfrak{R} \subset \mathfrak{N} + \mathfrak{R}, \quad \mathfrak{M} \times \mathfrak{R} \subset \mathfrak{N} \times \mathfrak{R}.$$

These relations, however, together with §4, Theorem 4, yield assertions β) and δ).

Rules β) and δ) cannot be sharpened, because, e.g.,

$$1 + \omega = 2 + \omega \quad \text{and} \quad 1 \cdot \omega = 2 \cdot \omega.$$

b) If $\mu < \nu$, $\rho < \sigma$, then $\mu + \rho < \nu + \sigma$; and if $0 < \mu < \nu$, $0 < \rho < \sigma$, then $\mu\rho < \nu\sigma$.

Proof: We have

$$\nu + \sigma \geq \mu + \sigma > \mu + \rho$$

and

$$\nu\sigma \geq \mu\sigma > \mu\rho.$$

The converses of rules a) obviously read as follows:

c) $\alpha)$ $\left\{\begin{matrix}\text{if } \rho + \mu < \rho + \nu \\ \text{or } \mu + \rho < \nu + \rho\end{matrix}\right\}$, then $\mu < \nu$;

$\beta)$ $\rho + \mu = \rho + \nu$ implies $\mu = \nu$;

$\gamma)$ $\left\{\begin{matrix}\text{if } \rho\mu < \rho\nu \\ \text{or } \mu\rho < \nu\rho\end{matrix}\right\}$, then $\mu < \nu$;

$\delta)$ if $\rho\mu = \rho\nu$, $\rho > 0$, then $\mu = \nu$.

For ordinal numbers it is also possible to define a subtraction and division, where, to be sure, essentially one-sided subtraction and division come into question, due to the absence of the commutative law for addition and multiplication.

Let $\mu < \nu$. Then, according to §4, Theorem 5, there is an ordinal number ζ such that $\mu + \zeta = \nu$, and, by c) β), there is only one such number ζ. This number can be designated as the difference $\zeta = -\mu + \nu$ of the two numbers ν and μ, where then, e. g., $-1 + \nu$ is to be distinguished from $\nu - 1$.

On the other hand, the equation $\xi + \mu = \nu$, with $\mu < \nu$, is not always solvable. For example, the equation $\xi + 1 = \omega$ cannot be solved, because the left-hand side represents an order type with a last element, whereas the right-hand side represents an order type without a last element.

The following result leads to division:

d) For any two ordinal numbers $\alpha > 0$ and μ, there is precisely one pair of ordinal numbers ξ and η such that

$$\mu = \alpha\eta + \xi, \quad \xi < \alpha.$$

Proof: Put $\beta = \mu + 1$. Then

$$\alpha\beta = \alpha(\mu + 1) \geq \mu + 1 > \mu.$$

Represent the ordinal numbers α, β by the sets $\mathfrak{A}$, $\mathfrak{B}$ with the elements a, b. Then $\alpha\beta$ is the order type of the ordered product of the second kind $\mathfrak{A} \times \mathfrak{B}$, i. e., of the set of pairs of elements (b, a) ordered according to first differences. This is a well-ordered set, because it represents an ordinal number. Since $\mu < \alpha\beta$, $\mathfrak{A} \times \mathfrak{B}$ contains a segment of order type μ. Let this segment be determined by the element (b_0, a_0). Under the order determined for $\mathfrak{A} \times \mathfrak{B}$, this segment consists of precisely those elements (y, x) for which $y \prec b_0$ in $\mathfrak{B}$ or $y = b_0$, $x \prec a_0$ in $\mathfrak{A}$. The element b_0 determines a segment in $\mathfrak{B}$, and hence, an ordinal number η; and a_0 determines a segment in $\mathfrak{A}$, and consequently, an ordinal number ξ. The set $\mathfrak{A} \times \mathfrak{B}$ can now be represented as an ordered sum of two terms, as follows: First come all elements (y, x) with $y \prec b_0$ and arbitrary x, then all elements of the form (b_0, x) with $x \prec a_0$. The first part is a set having ordinal number $\alpha\eta$, the second has ordinal number ξ. Thus $\mu = \alpha\eta + \xi$, with $\xi < \alpha$.

Now the numbers ξ, η are also uniquely determined. For let

$$\alpha\eta_1 + \xi_1 = \alpha\eta_2 + \xi_2 \,.$$

If $\eta_1 = \eta_2$, then $\xi_1 = \xi_2$ follows from c) β). But if $\eta_1 \neq \eta_2$, say $\eta_1 < \eta_2$, then $\eta_1 + 1 \leq \eta_2$, and the assumed equation therefore yields $\alpha\eta_1 + \xi_1 \geq \alpha\eta_1 + \alpha + \xi_2$. Hence, by c), $\xi_1 \geq \alpha + \xi_2$, which contradicts $\xi_1 < \alpha$. This second case, consequently, does not occur, and this establishes the uniqueness.

In the result just obtained, η plays the role of the quotient, and ξ, that of the remainder, in division. We can, of course, repeat the process, gaining thereby a Euclidean algorithm in the domain of ordinal numbers:

$$\alpha_0 = \alpha_1\eta_1 + \alpha_2\,, \qquad \alpha_2 < \alpha_1\,,$$

$$\alpha_1 = \alpha_2\eta_2 + \alpha_3\,, \qquad \alpha_3 < \alpha_2\,,$$

$$\alpha_2 = \alpha_3\eta_3 + \alpha_4\,, \qquad \alpha_4 < \alpha_3\,,$$

$$\cdots\cdots\cdots\cdots\,,$$

The process terminates after a finite number of steps. This is a consequence of Theorem 2 on p. 90, since the α_n's form a decreasing sequence. Hence, there is a natural number n such that $\alpha_n = 0$.

If, in d), we choose, in particular, $\alpha = \omega$, then for every ordinal number we get a representation

$$\mu = \omega\eta + \xi, \quad \xi < \omega,$$

i. e., one with a finite number as remainder. If μ is a limit number, then $\mu = \omega\eta$.

For $\alpha = 2$ it follows that every ordinal number is either of the form 2η or $2\eta + 1$, i. e., either "even" or "odd". For instance, ω is an even number, because (cf. p. 63) $\omega = 2\omega$. Every limit number ν is divisible by each of the numbers 1, 2, $\cdots$, ω; i. e., for every $0 < \alpha \leq \omega$, ν can be represented in the form $\nu = \alpha\eta$. For according to d), there is a representation

$$\nu = \alpha\eta + \xi, \quad \xi < \alpha.$$

If ξ here were >0, ν would be no limit number.

We take this opportunity to point out that the "set of all even ordinal numbers" is also a meaningless concept. For if this set had a meaning, there would exist an ordinal number ν greater than every even ordinal number. But then ν or $\nu + 1$ would be even and at the same time greater than every even ordinal number.

Concerning limit numbers in particular, there is the following rule:

e) Let ν be a limit number, and let α be any kind of ordinal number. Then $\alpha + \nu$ is also a limit number, and, if $\alpha \neq 0$, $\alpha\nu$ is a limit number too. In fact,

$$(1) \qquad \alpha + \lim_{\mu<\nu} \mu = \lim_{\mu<\nu} (\alpha + \mu), \qquad \alpha \cdot \lim_{\mu<\nu} \mu = \lim_{\mu<\nu} (\alpha\mu).$$

Proof: If the ordinal numbers $\alpha + \mu$ and $\alpha\mu$ are ordered according to increasing μ's, we obtain increasing sequences without a last element, and these sequences therefore define limit numbers.

Now let

$$\beta = \lim (\alpha + \mu).$$

According to the definition of ν, certainly $\alpha + \nu > \alpha +$ every μ. Hence, also $\alpha + \nu >$ every $(\alpha + \mu)$, so that

(2) $$\alpha + \nu \geq \beta.$$

Since, on the other hand, $\beta > \alpha$, there is a $\gamma > 0$ such that $\alpha + \gamma = \beta$. Then, by the definition of β,

$$\alpha + \gamma = \beta > \text{every } (\alpha + \mu),$$

so that

$$\alpha + \gamma > \alpha + \text{every } \mu,$$

and hence

$\gamma >$ every μ, and consequently $\gamma \geq \nu$; therefore $\alpha + \nu \leq \alpha + \gamma = \beta$. This, in connection with (2), yields the first of the relations (1).

To prove the second of these relations, put

$$\beta = \lim (\alpha\mu).$$

By the definition of ν, certainly $\alpha\nu > \alpha \cdot$(every μ). Hence, also

$$\alpha\nu > \text{every } (\alpha\mu), \text{ so that } \alpha\nu \geq \beta.$$

On the other hand, β can be represented in the form

$$\beta = \alpha\gamma + \delta, \quad \delta < \alpha.$$

Since $\beta >$ every $\alpha\mu$, it follows that

$$\text{every } \alpha\mu < \alpha\gamma + \delta < \alpha(\gamma + 1),$$

and therefore

$$\text{every } \mu < \gamma + 1.$$

Consequently, $\nu \leq \gamma + 1$, and hence, since ν is a limit number,

$$\nu < \gamma + 1, \text{ so that } \nu \leq \gamma;$$

therefore $\alpha\nu \leq \alpha\gamma \leq \beta$. This, in connection with $\alpha\nu \geq \beta$, proves the rest of the assertion.

7. *The Sequence of Ordinal Numbers, and Transfinite Induction*

We return once more to the considerations on p. 86; in particular, to Theorem 1 and the set $\mathfrak{W}_\mu$ consisting of the ordinal numbers which are less than μ. According to Theorem 1, every well-ordered set $\mathfrak{M}$ is similar to a set $\mathfrak{W}_\mu$. This means that the elements m of the set $\mathfrak{M}$ can be associated with the ordinal numbers $\rho < \mu$ in such a one-to-one manner, that, if m_1, m_2 are any two elements, $m_1 \prec m_2$ if, and only if, for the corresponding ordinal numbers ρ_1, ρ_2, we have also $\rho_1 < \rho_2$.

In other words—this will make the fact seem more familiar—we can count off the elements of every well-ordered set of ordinal number μ with the aid of the ordinal numbers $<\mu$, a fact which of course is trivial for sets with a finite number of elements. For example, the set $\{a, b, c, d\}$ having ordinal number 4 can be counted with the ordinal numbers 0, 1, 2, 3. If we keep this counting process for finite sets in mind, it seems hopeless at first to obtain an analogous counting process for infinite sets. All the more highly must we value the accomplishment of Cantor's, who, by the introduction of well-ordered sets and ordinal numbers, overcame these difficulties.

It will be well to note the first numbers of a set $\mathfrak{W}_\mu$ for sufficiently large μ.

First come the finite numbers, ordered according to increasing magnitude. After the set of all finite ordinal numbers we find the first transfinite ordinal number. It is, according to Theorem 1 on p. 86, precisely the order type of the set of all ordinal numbers preceding it; in this case, therefore, the order type of the set $\{0, 1, 2, \cdots\}$; i. e., ω. This ordinal number is essentially different from those preceding it. For, ω has no immediate predecessor; it is a limit number.

To ω, as to every ordinal number, there corresponds an immediate successor $\omega + 1$; then comes $\omega + 2$; etc. We are thus led to the sequence of ordinal numbers

$$0, 1, 2, \cdots, \omega, \omega + 1, \omega + 2, \cdots.$$

Since this sequence has order type $\omega + \omega = \omega \cdot 2$, the number coming next after this sequence is $\omega \cdot 2$. This ordinal number is followed by $\omega \cdot 2 + 1, \omega \cdot 2 + 2, \cdots$, and after all these comes $\omega \cdot 3$; etc. Thus the beginning of the sequence of ordinal numbers is

$$0, 1, 2, \cdots, \quad \omega, \omega + 1, \omega + 2, \cdots, \quad \omega \cdot 2, \omega \cdot 2 + 1, \cdots,$$

$$\omega \cdot n, \omega \cdot n + 1, \cdots .$$

This set has order type ω^2. Therefore the next ordinal number is ω^2. This is succeeded by $\omega^2 + 1, \omega^2 + 2, \cdots, \omega^2 + \omega, \cdots$, more generally, all ordinal numbers of the form $\omega^2 + \omega \cdot n_1 + n_0$, where n_0 and n_1 are finite ordinal numbers. Now the set of all the ordinal numbers considered thus far consists of two sequences of order type ω^2 placed one after the other, and therefore as a whole it represents the ordinal number $\omega^2 \cdot 2$, so that this is the next ordinal number. If we continue in this manner, we obtain all ordinal numbers which can be written in the form of "polynomials"

$$\omega^k \cdot n_k + \omega^{k-1} \cdot n_{k-1} + \cdots + \omega \cdot n_1 + n_0 \tag{1}$$

with finite numbers k and n_i. The ordinal number which appears after all these numbers (1) can no longer be expressed in terms of ω, at least not until products with infinitely many factors or arbitrary powers of ordinal numbers have been introduced, which will take place in the next paragraph. This will require transfinite induction, which we shall discuss now.

Suppose that we wish to prove let us say the binomial formula

$$(a + b)^n = \sum_{k=0}^{n} \binom{n}{k} a^{n-k} b^k$$

for natural numbers n, or the general validity of an assertion $A(n)$ set up for every natural number n, with the aid of "complete (or mathematical) induction" or "inference from n to $n + 1$". Then, as is well-known, we proceed as follows:

First, we prove the validity of the proposition $A(n)$ for the smallest number for which the truth of $A(n)$ is affirmed, say for $n = 0$.

Second, we prove that *if* the proposition $A(n)$ holds for all numbers $n < n_0$, it is also true for n_0.

Third, from this now follows the universal validity of the proposition $A(n)$. For if $A(n)$ were false for some n, there would be a smallest such number, call it n_0. On account of the first step, $n_0 > 0$. $A(n)$, then, is true for the n which precedes n_0, and consequently, by the second step, also for n_0 itself. The assumption "$A(n)$ is false" has thus proved to be untenable.

Corresponding to this method of proof we can now also proceed with assertions $A(\mu)$ which refer to an arbitrary ordinal number μ.

First, we prove the validity of $A(\mu)$ for the smallest ordinal number μ that comes into question, say $\mu = 0$.

Second, we prove that *if* the proposition is true for all ordinal numbers $\mu < \mu_0$, then it is also true for μ_0.

Third, from this follows the universal validity of the proposition $A(\mu)$. For suppose that $A(\mu)$ were false for some $\bar{\mu}$. Consider the well-ordered set $\mathfrak{W}_{\bar{\mu}+1}$, and in this take the subset of numbers μ for which $A(\mu)$ is false. This set, as a subset of a well-ordered set, has a first element μ_0 which, on account of the first step, is >0. The set of numbers $\mu < \mu_0$ is therefore not empty. According to the definition of μ_0, $A(\mu)$ is true for all numbers $\mu < \mu_0$, and hence, by the second step, also for μ_0 itself. The assumption "$A(\mu_0)$ is false" has thus proved to be untenable. Consequently, by the logical law of the excluded middle, the assertion $A(\mu)$ is universally valid.

This procedure is frequently called transfinite induction, because it is now no longer restricted to finite sets. It can be used for definitions as well as for proofs. Examples of this are contained in the next paragraph.

8. *The Product of Arbitrarily Many Ordinal Numbers*

Up to now we have defined multiplication of only a finite number of ordinal numbers, in particular, of two ordinal numbers. The definition of a product of arbitrarily many factors will take place in the same way that we proceeded from a product of two factors to one of finitely many factors, viz., by the successive performance of multiplications of two factors, where now, of course, there will be an additional limit process. First, however, the analogue will take place for the sum, notwithstanding the fact that the sum of arbitrarily many terms has already been defined. On the basis of the earlier definition of sum, we shall derive two properties of sums; and then we shall show that, conversely, the sum can already be defined by means of these two properties.

Let a well-ordered complex $\mathfrak{K}(\mu)$ of ordinal numbers be given, where $\mathfrak{K}$ has ordinal number κ. By Theorem 1 on p. 86, the set $\mathfrak{K}$ is similar to the set $\mathfrak{W}_\kappa$, if the elements $\rho < \kappa$ of the latter set are ordered according to increasing magnitude. The given complex of ordinal numbers can therefore also be defined so that to every ordinal number $\rho < \kappa$ there corresponds an ordinal number μ_ρ.

We now keep fixed the ordinal number κ as well as the $\mu_\rho > 0$ associated with each $\rho < \kappa$. We then consider all sums $\sum_{\rho<\lambda} \mu_\rho$, where the terms are ordered according to increasing magnitude of the ρ's, and we prove:

(I) If $0 < \lambda \leq \kappa$, and if λ is not a limit number, then

$$\sum_{\rho<\lambda-1} \mu_\rho + \mu_{\lambda-1} = \sum_{\rho<\lambda} \mu_\rho .$$

(II) If $0 < \lambda \leq \kappa$, and if λ is a limit number, then $\sum_{\rho<\lambda} \mu_\rho$ is a limit number too, and

$$\lim_{\sigma<\lambda} \sum_{\rho<\sigma} \mu_\rho = \sum_{\rho<\lambda} \mu_\rho .$$

Relation (I) follows immediately from Definition 4 on p. 61, if the μ_ρ's appearing in (I) are represented by mutually exclusive sets.

To prove (II), represent the μ_ρ's likewise by well-ordered disjunct sets $\mathfrak{M}_\rho$. It is clear that the sums $\sum_{\rho<\sigma} \mu_\rho$, ordered according to increasing σ for all $\sigma < \lambda$, form an increasing sequence, and, indeed, one without a last element, because λ is a limit number. Hence, in any case,

$$\lim_{\sigma<\lambda} \sum_{\rho<\sigma} \mu_\rho \leq \sum_{\rho<\lambda} \mu_\rho .$$

It now remains to be shown that there is no ordinal number between the left- and right-hand members of this inequality; in other words, that, for every $\alpha < \sum_{\rho<\lambda} \mu_\rho$, also $\alpha < \lim_{\sigma<\lambda} \sum_{\rho<\sigma} \mu_\rho$. Let, then, $\alpha < \sum_{\rho<\lambda} \mu_\rho$. Then α is the ordinal number of a segment of $\sum_{\rho<\lambda} \mathfrak{M}_\rho$. This segment is determined by an element which belongs to the set $\mathfrak{M}_\tau$, say. Since λ is a limit number, we have also $\tau + 1 < \lambda$ and α is the ordinal number of a segment of $\sum_{\rho<\tau+1} \mathfrak{M}_\rho$. Consequently,

$$\alpha < \sum_{\rho<\tau+1} \mu_\rho < \lim_{\sigma<\lambda} \sum_{\rho<\sigma} \mu_\rho ,$$

which proves the assertion.

With the help of properties (I) and (II), we can now also define the sum of arbitrarily many ordinal numbers, by means of transfinite induction, as follows, if the sum of two ordinal numbers has been defined. We put

$$(\alpha) \qquad f(0) = 0;$$

further, if $f(\rho)$ has already been defined for all $\rho < \lambda$:

$$(\beta) \qquad f(\lambda) = f(\lambda - 1) + \mu_{\lambda-1}$$

if λ is not a limit number, and

$$(\gamma) \qquad f(\lambda) = \lim_{\rho<\lambda} f(\rho)$$

if λ is a limit number.

Herewith $f(\lambda)$ is defined for every ordinal number $\lambda \leq \kappa$, and, by what has preceded, we have precisely

$$f(\lambda) = \sum_{\rho<\lambda} \mu_\rho .$$

This definition is obviously nothing but, e. g., the sum definition

$$a + b + c + d = \{((a + b) + c) + d\}$$

carried over to arbitrary well-ordered complexes of terms.

According to the pattern just carried out, a product of an arbitrary well-ordered complex of ordinal numbers is now defined as follows: Let the complex be defined again so that to every $\lambda < \kappa$ there corresponds an ordinal number μ_λ and the μ_λ's are ordered according to increasing magnitude of the λ's. The definition of a product of two factors is assumed to be known. Keeping fixed the numbers introduced so far, the product $\prod_{\rho<\lambda} \mu_\rho$ is defined for all $\lambda \leq \kappa$ as follows: Put

(a) $f(0) = 1$;

further, if $f(\rho)$ is known already for all $\rho < \lambda$,

(b) $f(\lambda) = f(\lambda - 1) \cdot \mu_{\lambda-1}$ if λ is not a limit number, and

(c) $f(\lambda) = \lim_{\rho<\lambda} f(\rho)$ if λ is a limit number.

Then, for all $\lambda \leq \kappa$, $f(\lambda)$ is defined inductively, and shall be the product

$$\prod_{\rho<\lambda} \mu_\rho = f(\lambda).$$

It is obvious that

$$f(1) = \mu_0 \, , f(2) = \mu_0 \cdot \mu_1 \, , \cdots , f(n) = \mu_0 \cdot \mu_1 \cdots \mu_{n-1} \, ,$$

so that this definition coincides with the one given earlier for a finite product.

We shall not derive rules of operation for the general product here, because we shall not need them. We must, however, point out an important difference between the two kinds of product. If we form, e. g., the infinite product $2 \cdot 2 \cdot 2 \cdots$ with enumerably many factors, this product, by our definition, is equal to

$$\lim_{n<\omega} 2^n = \text{the next number after } \{2^0, 2^1, 2^2, \cdots\},$$

i. e., it equals ω. In this connection, the 2's here are to be interpreted as ordinal numbers. But if we interpret the product as a product of cardinal numbers, we obtain $2^a = \mathfrak{c}$. Thus, whereas for finite products of ordinal numbers we had invariably $|\mu \cdot \nu| = |\mu| \cdot |\nu|$, this is by no means the case here, since

$|\omega| = \mathfrak{a} < \mathfrak{c}$. The same is true for the product $2 \cdot 3 \cdot 4 \cdots$. As a product of ordinal numbers, it is

$$\lim_{n<\omega} n! = \omega;$$

as a product of cardinal numbers,

$$2 \cdot 3 \cdot 4 \cdots \geq 2^{\mathfrak{a}} = \mathfrak{c}.$$

We must therefore note that for infinite products of ordinal numbers it is possible to have

$$\left| \prod_{\rho<\kappa} \mu_\rho \right| \neq \prod_{\rho<\kappa} |\mu_\rho|.$$

This is due to the fact that products of infinitely many ordinal numbers are, to be sure, again ordinal numbers, on the basis of the definition of such products, but they are no longer defined as order types of products of sets.

9. *Powers of Ordinal Numbers*

The power μ^α is now once more defined as a product of equal factors. By specializing the general definition of the product, we obtain the following definition of the power: Put
(a) $f(0) = 1$;
further, if $f(\rho)$ is already defined for all $\rho < \lambda$,
(b) $f(\lambda) = f(\lambda - 1) \cdot \mu$ if λ is not a limit number, and
(c) $f(\lambda) = \lim_{\rho<\lambda} f(\rho)$ if λ is a limit number.
Then we stipulate that $\mu^\lambda = f(\lambda)$.

For example, according to this we have

$$2^\omega = 2 \cdot 2 \cdot 2 \cdots = \omega; \qquad \omega^\omega = \omega \cdot \omega \cdots = \lim_{n<\omega} \omega^n.$$

Moreover,

$$1 + \omega = \omega,$$

$$1 + \omega + \omega^2 = \omega + \omega^2 = \omega(1 + \omega) = \omega^2,$$

$$1 + \omega + \omega^2 + \omega^3 = \omega^2 + \omega^3 = \omega^2(1 + \omega) = \omega^3,$$

$$\cdots\cdots\cdots\cdots\cdots\cdots\cdots\cdots,$$

Consequently, by p. 101 (II), also

$$1 + \omega + \omega^2 + \omega^3 + \cdots = \omega^\omega.$$

We can now continue the investigation on p. 99. There we left open the question of how to denote the ordinal number which comes after all polynomials

$$(1) \qquad \zeta_k = \omega^k n_k + \omega^{k-1} n_{k-1} + \cdots + \omega n_1 + n_0 \, .$$

For every ζ_k, certainly $\omega^k \leq \zeta_k < \omega^{k+1}$. Hence, by p. 91, Theorem 5, the fundamental sequence of the ζ_k's has the same limit as the fundamental sequence $1, \omega, \omega^2, \omega^3, \cdots$, viz., the limit ω^ω. Thus, after the polynomial (1) come the ordinal numbers $\omega^\omega, \omega^\omega + 1, \cdots, \omega^\omega \cdot \omega^n, \cdots$.

Since the commutative law for the product does not hold, we lose one of the usual laws of operation for powers. For it can happen that $(\mu\nu)^\rho \neq \mu^\rho \nu^\rho$, as is shown, e. g., by $\nu = \rho = 2$, $\mu = \omega$. We do have, however,

a) $\mu^\alpha \mu^\beta = \mu^{\alpha+\beta}$,

b) $(\mu^\alpha)^\beta = \mu^{\alpha\beta}$.

Proof[2]: For fixed μ and α, the rules are proved by induction for every β. For $\beta = 0$ they are trivially correct. Let us assume, then, that they have already been verified for all $\gamma < \beta$. If β is not a limit number, then, by the definition of the power, and the rule which we have assumed is true for $\beta - 1$,

$$\mu^\alpha \cdot \mu^\beta = \mu^\alpha(\mu^{\beta-1} \cdot \mu) = (\mu^\alpha \mu^{\beta-1}) \cdot \mu$$

$$= \mu^{\alpha+\beta-1} \cdot \mu = \mu^{\alpha+\beta};$$

$$(\mu^\alpha)^\beta = (\mu^\alpha)^{\beta-1} \cdot \mu^\alpha = \mu^{\alpha(\beta-1)} \cdot \mu^\alpha = \mu^{\alpha(\beta-1)+\alpha} = \mu^{\alpha\beta}.$$

If β is a limit number, $\beta = \lim \gamma$, then, bearing in mind p. 96, e),

---

[2]Since a) is used to prove b), we must first carry out the proof of a) and then that of b). In the text, however, both proofs are given together in order to save space,

$$\mu^\alpha \cdot \mu^\beta = \mu^\alpha \cdot \lim \mu^\gamma = \lim (\mu^\alpha \mu^\gamma) = \lim \mu^{\alpha+\gamma}$$

$$= \mu^{\lim(\alpha+\gamma)} \quad = \mu^{\alpha+\lim\gamma} \quad = \mu^{\alpha+\beta};$$

$$(\mu^\alpha)^\beta = \lim (\mu^\alpha)^\gamma \quad = \lim \mu^{\alpha\gamma} \quad = \mu^{\lim \alpha\gamma}$$

$$= \mu^{\alpha \cdot \lim\gamma} \quad = \mu^{\alpha\beta}.$$

Thus, e.g.,

$$(\omega^\omega)^\omega = \omega^{\omega^2}, \qquad (\omega^{\omega^2})^\omega = \omega^{\omega^3}.$$

In the fundamental sequence

$$\omega^\omega, \omega^{\omega^2}, \omega^{\omega^3}, \cdots,$$

the exponents form a fundamental sequence having the limit ω^ω. Therefore the above fundamental sequence has the limit

$$\omega^{\omega^\omega} = \omega^{(\omega^\omega)}.$$

We can continue in this manner, obtaining finally the fundamental sequence

$$1, \omega, \omega^\omega, \omega^{\omega^\omega}, \omega^{\omega^{\omega^\omega}}, \cdots,$$

whose limit will be denoted by ϵ_0. According to the definition of the power,

$$\omega^{\epsilon_0} = \lim \{\omega, \omega^\omega, \omega^{\omega^\omega}, \cdots\} = \epsilon_0 .$$

Cantor called every solution of the equation $\omega^\epsilon = \epsilon$ an ϵ-number.

As to inequalities for powers, we mention the following:

c) If $\mu \leq \nu$ and $\alpha \geq 0$, then $\mu^\alpha \leq \nu^\alpha$.

Proof: The assertion is true for $\alpha = 0$. Suppose that it is true already for all $\gamma < \alpha$. Then, if α is not a limit number, it follows from p. 105, a) that

$$\mu^\alpha = \mu^{\alpha-1} \cdot \mu \leq \nu^{\alpha-1} \cdot \mu \leq \nu^{\alpha-1} \cdot \nu = \nu^\alpha;$$

and if α is a limit number, $\alpha = \lim \gamma$, it follows from p. 91, Theorem 5 that

$$\mu^\alpha = \lim \mu^\gamma \leq \lim \nu^\gamma = \nu^\alpha,$$

d) If $\alpha < \beta$ and $\mu > 1$, then $\mu^\alpha < \mu^\beta$.

Proof: The assumption implies the existence of a number $\gamma > 0$ such that $\alpha + \gamma = \beta$. But then

$$\mu^\beta = \mu^\alpha\mu^\gamma > \mu^\alpha.$$

e) For $\mu > 1$ and $\alpha \geq 1$, $\mu^\alpha \geq \alpha$.

Proof: The assertion is true for $\alpha = 1$. Let it be true already for all $1 \leq \beta < \alpha$. Then it is also true for α. For suppose that α is not a limit number. Then, on account of $\alpha \geq 2$,

$$\mu^\alpha = \mu^{\alpha-1}\cdot\mu \geq (\alpha - 1)\mu \geq (\alpha - 1)\cdot 2$$

$$= (\alpha - 1) + (\alpha - 1) \geq (\alpha - 1) + 1 = \alpha.$$

If, however, α is a limit number, $\alpha = \lim \beta$, then

$$\mu^\alpha = \lim \mu^\beta \geq \lim \beta = \alpha.$$

10. Polynomials in Ordinal Numbers

Ordinal numbers exhibit a behavior similar to that of the natural numbers in several other respects. Just as every natural number can be represented as a decimal, i. e., in terms of powers of 10, more generally, in terms of powers of an arbitrary natural number $b > 1$, for ordinal numbers we have

THEOREM 1. *Let $\beta > 1$ be an arbitrary ordinal number. Then every ordinal number $\zeta > 0$ can be represented in one, and only one, way in the form*

$$\zeta = \beta^\alpha\gamma + \beta^{\alpha_1}\gamma_1 + \cdots + \beta^{\alpha_n}\gamma_n, \tag{1}$$

with $\alpha > \alpha_1 > \cdots > \alpha_n \geq 0$ and $0 < \gamma, \gamma_1, \cdots, \gamma_n < \beta$; i. e., as a finite sum of powers with coefficients which are smaller than the base.

Proof: The existence of the representation is established as follows: According to §9, e), $\beta^{\zeta+1} > \zeta$. Hence, there is a smallest number δ such that $\beta^\delta > \zeta$. Here δ is not a limit number, because otherwise, for every $\eta < \delta$, we should have successively

$\eta + 1 < \delta$, $\beta^{\eta+1} \leq \zeta$, $\beta^{\eta} < \zeta$, and finally $\beta^{\delta} = \lim \beta^{\eta} \leq \zeta$, which contradicts $\beta^{\delta} > \zeta$.

Since δ is not a limit number, it has an immediate predecessor $\alpha = \delta - 1$, and by the definition of δ,

$$\beta^{\alpha} \leq \zeta < \beta^{\alpha+1}. \tag{2}$$

According to p. 94, d), there exist numbers γ, ζ_1 such that

$$\zeta = \beta^{\alpha}\gamma + \zeta_1, \quad \zeta_1 < \beta^{\alpha}.$$

Because of (2), it follows that $0 < \beta^{\alpha}\gamma < \beta^{\alpha+1} = \beta^{\alpha}\beta$, and hence $0 < \gamma < \beta$. The process can be repeated with ζ_1, provided that $\zeta_1 > 0$. We thus obtain a chain of equations

$$\begin{cases} \zeta = \beta^{\alpha}\gamma + \zeta_1, \\ \zeta_1 = \beta^{\alpha_1}\gamma_1 + \zeta_2, \\ \cdot \quad \cdot \quad \cdot \quad \cdot \quad \cdot \quad \cdot \quad \cdot \end{cases} \tag{3}$$

with the inequalities

$$\zeta \geq \beta^{\alpha} > \zeta_1 \geq \beta^{\alpha_1} > \zeta_2 \geq \cdots,$$

from which follows $\alpha > \alpha_1 > \alpha_2 > \cdots$ as well as $0 < \gamma, \gamma_1, \cdots < \beta$. Since a decreasing sequence contains only a finite number of terms, the process terminates after a finite number of steps. This means that there exists a $\zeta_{n+1} = 0$. The representation then follows from equations (3).

The uniqueness of the representation is proved as follows: For every number ζ of the form (1), it is easy to see that $\zeta < \beta^{\alpha+1}$. If we had

$$\zeta = \beta^{\alpha}\gamma + \beta^{\alpha_1}\gamma_1 + \cdots = \beta^{\bar{\alpha}}\bar{\gamma} + \beta^{\bar{\alpha}_1}\bar{\gamma}_1 + \cdots \tag{4}$$

and $\alpha < \bar{\alpha}$, then $\alpha + 1 \leq \bar{\alpha}$, and (4) would imply

$$\beta^{\alpha+1} > \zeta \geq \beta^{\bar{\alpha}} \geq \beta^{\alpha+1},$$

which is impossible. Therefore $\bar{\alpha} = \alpha$. If we now had $\bar{\gamma} > \gamma$, say $\bar{\gamma} = \gamma + \bar{\bar{\gamma}}$, then a leading term would drop out in (4), giving

$$\beta^{\alpha_1}\gamma_1 + \cdots = \beta^{\alpha}\bar{\bar{\gamma}} + \cdots ,$$

which is impossible again, because $\alpha_1 < \alpha$.

The theorem says, e. g., for $\beta = 2$, that every ordinal number ζ can be represented in the form

$$\zeta = 2^{\alpha} + 2^{\alpha_1} + \cdots + 2^{\alpha_n}.$$

Moreover (p. 104), $\omega = 2^{\omega}$. In such a representation, therefore, the highest exponent that occurs has by no means to be smaller than the number represented.

We obtain an especially important example for $\beta = \omega$. It follows then that every ordinal number ξ can be represented in the form

$$\xi = \omega^{\alpha}c + \omega^{\alpha_1}c_1 + \cdots + \omega^{\alpha_n}c_n ,$$

where the c_n's are less than ω, so that they are finite numbers. If we add terms with zero coefficients, we obtain for every ordinal number ξ a representation

$$\xi = \cdots + \omega^{\omega}x_{\omega} + \cdots + \omega^2 x_2 + \omega x_1 + x_0$$

$$= {}^*\sum_{\alpha} \omega^{\alpha}x_{\alpha} ,$$

where the star before the $\sum$ indicates that this sum is ordered according to decreasing powers of the base ω, and where the x_{α}'s are integers ≥ 0. This representation only apparently contains infinitely many terms. Actually only a finite number of nonzero terms occur. If we agree that the representation shall begin with a nonzero term, then the representation is uniquely determined by ξ. Now let ξ, η be two arbitrary ordinal numbers, having the representations

$$\xi = {}^*\sum_{\alpha} \omega^{\alpha}x_{\alpha} , \qquad \eta = {}^*\sum_{\alpha} \omega^{\alpha}y_{\alpha} .$$

Then we can form their so-called "natural sum" (G. Hessenberg)

$$\sigma(\xi, \eta) = \sum_{\alpha} \omega^{\alpha}(x_{\alpha} + y_{\alpha})$$

which also contains only a finite number of nonzero coefficients.

Since the natural sum is a sum of ordinal numbers, it certainly represents an ordinal number; and obviously $\sigma(\xi, \eta) = \sigma(\eta, \xi)$ because the commutative law of addition holds for natural numbers. But this natural sum has nothing to do with the sum $\xi + \eta$ of the two ordinal numbers ξ, η. For example, for $\xi = \omega$, $\eta = \omega^2 + 1$, we have

$$\xi + \eta = \omega + \omega^2 + 1 = \omega(1 + \omega) + 1 = \omega^2 + 1,$$

$$\eta + \xi = \omega^2 + 1 + \omega = \omega^2 + \omega,$$

$$\sigma(\xi, \eta) = \omega^2 + \omega + 1,$$

and these three results are different from one another.

In analogous fashion, by formal multiplication, one can also form natural products.

Later on we shall need the following property of the natural sum:

THEOREM 2. *For a given ordinal number ζ, the equation*

$$\sigma(\xi, \eta) = \zeta \tag{5}$$

has only a finite number of solutions ξ, η.

Proof: The number ζ can be represented in the form

$$\zeta = {}^*\sum_{\alpha} \omega^{\alpha} z_{\alpha} .$$

If equation (5) is to hold, we must have

$$z_{\alpha} = x_{\alpha} + y_{\alpha}$$

for every α. This equation, for every α, has exactly $1 + z_{\alpha}$ solutions. Since, however, there are but finitely many $z_{\alpha} \neq 0$, there exist only finitely many solution systems x_{α} and y_{α}, and hence, only a finite number of solutions ξ, η.

11. The Well-ordering Theorem

Earlier already we raised the question whether every set can be well-ordered. We have already seen that an affirmative answer to this question is also of the greatest importance for the theory of cardinal numbers. It would show, at the same time, that every set can be represented as an ordered set. This question too was left open.

Cantor regarded the possibility of well-ordering every set as a logical necessity. The following argument can be given for this opinion: If a set $\mathfrak{M}$ is given, pick out an element m_0 and put it in first position. Then choose an arbitrary element m_1 in $\mathfrak{M} - \{m_0\}$, and put it in second position. Then select in any manner an element m_2 in $\mathfrak{M} - \{m_0, m_1\}$, and assign it the next place. Continue in this fashion. If the process terminated in any manner before the set $\mathfrak{M}$ was entirely exhausted, we would still have a nonempty subset of $\mathfrak{M}$ left, and could again pick out an element of this subset and place it after all the elements already chosen. The process would thus continue nevertheless, contrary to our assumption; it can therefore come to an end only when the set $\mathfrak{M}$ has been used up entirely.

This argument, of course, merely makes plausible the possibility of well-ordering every set. Think only of the complicated structure which a well-ordered set can have, in particular, of the many successive limit numbers, of which there need by no means be only an enumerable number.

An actual proof of the well-ordering theorem was first given by E. Zermelo (Mathematische Annalen, vol. 59(1904), pp. 514–516; a second proof in Mathematische Annalen, vol. 65(1908), pp. 107–128). In the first of the two proofs, the significance of the individual steps can be followed better than in the second proof, and we shall therefore reproduce the first of the two proofs here. In the proof, we do not first well-order the whole set, but rather proceed from well-ordered subsets of the set $\mathfrak{M}$ which is to be well-ordered. Such subsets can be constructed according to the simple procedure just described, and we certainly arrive at finite and enumerable well-ordered subsets of $\mathfrak{M}$. To be able to get beyond these well-ordered subsets, they are, to be sure, subjected to an additional condition. Starting from these special well-ordered subsets, it is then possible to well-order the entire set. The selection of elements from $\mathfrak{M}$

plays a role in the proof. Now, however, the selections are not made successively, as in the argument presented before; on the contrary, at the start an element is chosen from every nonempty subset of $\mathfrak{M}$. How this is to take place remains undecided. Having made these preparatory remarks, we now prove the

WELL-ORDERING THEOREM. *Every set $\mathfrak{M}$ can be well-ordered.*

Proof: That every finite set can be well-ordered has already been mentioned. We therefore assume that $\mathfrak{M}$ is an infinite set, and we break up the proof into a series of steps.

I. Let the elements of $\mathfrak{M}$ be denoted by m, and let $\mathfrak{U}(\mathfrak{M})$ be the set of all subsets of $\mathfrak{M}$. In every nonempty subset $\mathfrak{N}$ of $\mathfrak{M}$ we choose in any manner an element n and associate it with the set $\mathfrak{N}$. We call this element the "distinguished" element of $\mathfrak{N}$ and denote it also by

$$n = \varphi(\mathfrak{N}).$$

Here it is by no means necessary that distinct subsets $\mathfrak{N}$ have distinct distinguished elements. This correspondence between elements n and the subsets $\mathfrak{N}$ is also designated as a *covering* of the set $\mathfrak{U}(\mathfrak{M})$, and remains fixed throughout the entire proof.

II. For the set $\mathfrak{M}$ and the established covering of $\mathfrak{U}(\mathfrak{M})$ with distinguished elements, a nonempty well-ordered subset Γ of $\mathfrak{M}$ shall be called a Γ-sequence, if for every element c of Γ and for the segment Γ_c of Γ determined by c (Γ_c may be 0, but invariably $\mathfrak{M} - \Gamma_c \neq 0$) the relation

$$\varphi(\mathfrak{M} - \Gamma_c) = c$$

holds.

By means of this definition, the auxiliary condition for well-ordered sets announced at the beginning is introduced. If we have *a* Γ-sequence, we now no longer need to choose any element from $\mathfrak{M} - \Gamma_c$; such an element is already given with Γ_c, viz., c.

Γ-sequences always exist. For if, e. g., $m_0 = \varphi(\mathfrak{M})$, then obviously $\{m_0\}$ is a Γ-sequence. But we can say even more: Every Γ-sequence has m_0 as first element. This is true because for the first element c of a Γ-sequence we have $\Gamma_c = 0$, and hence $c = \varphi(\mathfrak{M}) = m_0$.

Further, if m_1 is the distinguished element of $\mathfrak{M} - \{m_0\}$, then obviously $\{m_0, m_1\}$ is a Γ-sequence. Moreover, every Γ-sequence with at least two elements has m_1 as second element. For if c is the second element, then, since the first element is m_0,

$$\Gamma_c = \{m_0\}, \text{ and hence } c = \varphi(\mathfrak{M} - \{m_0\}) = m_1 .$$

For the two smallest Γ-sequences we have thus established the fact that they are segments of every Γ-sequence which contains at least the same number of elements. We have, indeed, quite generally, the following result:

III. Of two distinct Γ-sequences it is invariably the case that one is a segment of the other.

For let Γ, Γ^{**} be two distinct Γ-sequences. Then, in any case, since well-ordered sets are always comparable, one of them, say Γ, is similar to a segment Γ^* of the other. We have now to show that $\Gamma \simeq \Gamma^*$ here implies actually $\Gamma = \Gamma^*$. In other words, if the element of Γ^* associated with an element $c \in \Gamma$ is denoted by c^*, we have to prove that invariably $c = c^*$. That this is true for the first element of Γ has already been shown in II. It follows in general by means of transfinite induction. Suppose that the assertion is correct for all $c \prec c_1$. Then the segments Γ_{c_1} and $\Gamma^*_{c_1}$ coincide. Then, since

$$c_1 = \varphi(\mathfrak{M} - \Gamma_{c_1}), \qquad c_1^* = \varphi(\mathfrak{M} - \Gamma^*_{c_1}),$$

also $c_1 = c_1^*$, which proves the assertion.

From this immediately follows:

IV. If two Γ-sequences have an element c in common, then their segments determined by c coincide too. And from this we get the following: If two Γ-sequences have the two elements a and b in common, then either $a \prec b$ in *both* sequences or $a \succ b$ in *both* sequences.

V. The union Σ of all Γ-sequences can be ordered.

For if a, b are two distinct elements of Σ, they belong to two Γ-sequences, say Γ and Γ^*. If $\Gamma \neq \Gamma^*$, then, according to III, one of them, say Γ, is a segment of the other. Both elements a, b then belong to Γ^*. By virtue of the order of the elements in Γ^*, one of the order relations $a \prec b$ or $a \succ b$ is determined for a, b. By the second part of IV, this determination is independent of the particular set Γ^* to which the elements belong. This order relation shall now hold for a, b in Σ too. This determines an order for Σ, provided that we can show that the order relation is transitive. Let

$$a \prec b, \quad b \prec c$$

for three elements of Σ. Let Γ determine the order of a, b, and Γ^* the order of b, c. Then, by IV, Γ^* also contains the segment Γ_b, and hence, in particular, the element a. Since the order of two elements is determined by *every* Γ-sequence containing them, that of a, c, in particular, is also determined by Γ^*, i. e., $a \prec c$, as was asserted.

VI. The set Σ is actually well-ordered by the order just determined.

We have to show that every nonempty subset Σ^* of Σ has a first element. Let c^* be an arbitrary element of Σ^*, where for the proof we may assume that c^* is not already the first element of Σ^*. The element c^* belongs to a set Γ^*. Let a^* be an arbitrary element which precedes c^* in Σ^*. Then, according to the determination of the order in Σ, a^* precedes c^* in a certain Γ-sequence, and hence, by IV, also in Γ^*. The elements which precede c^* in Σ^* thus form a subset of the well-ordered set Γ^*, and there is therefore a first element among them, Q.E.D.

VII. The set Σ is actually a Γ-sequence.

For let Σ_c be the segment determined by an element c of Σ. The element c belongs to a set Γ, and determines in this set a segment Γ_c. According to the argument presented in VI, every element of Σ_c belongs to Γ_c, and the converse follows from the ordering of Σ. Hence, $\Sigma_c = \Gamma_c$, and consequently

$$\varphi(\mathfrak{M} - \Sigma_c) = \varphi(\mathfrak{M} - \Gamma_c) = c, \quad \text{Q.E.D.}$$

VIII. $\mathfrak{M} = \Sigma$, and hence $\mathfrak{M}$ is well-ordered by the given procedure.

For if we had $\Sigma \subset \mathfrak{M}$, $\mathfrak{M} - \Sigma$ would be a nonempty subset of $\mathfrak{M}$, and would possess a distinguished element z. Then $\Sigma + \{z\}$ would also be a well-ordered set; here the sum is meant to be an ordered sum, so that z comes after all the elements of Σ. For every segment Γ_c of $\Sigma + \{z\}$, determined by an element c of this set,

$$\varphi(\mathfrak{M} - \Gamma_c) = c;$$

because for $c \neq z$ this is true by VII, and for $c = z$, by the definition of z. Thus $\Sigma + \{z\}$ would also be a Γ-sequence, contradicting the fact that Σ, by definition, contains every Γ-sequence.

From the well-ordering theorem it follows, e. g., that the continuum can be well-ordered, although it has been impossible thus far to produce a specific well-ordering of the continuum. The well-ordering theorem is just a pure existence-theorem.

With regard to subsequent critical remarks, note that, starting from the given set $\mathfrak{M}$, only the following sets were newly constructed: subsets of $\mathfrak{M}$, the power set $\mathfrak{U}(\mathfrak{M})$, and subsets of $\mathfrak{U}(\mathfrak{M})$. The formation of supersets of $\mathfrak{M}$ was thus restricted to the formation of the power set.

12. An Application of the Well-ordering Theorem

Before we discuss the effects of the well-ordering theorem on the theory of sets, let us treat an interesting application of this theorem, which goes back to G. Hamel (Mathematische Annalen, vol. 60 (1905), pp. 459–462).

Cauchy considered the problem of determining a function $f(x)$ such that

$$(1) \qquad f(x + y) = f(x) + f(y).$$

The function cx obviously has this property, for every constant c; and if only continuous functions are sought, there are no other solutions of (1). For, put $f(1) = c$. Then from (1) we get

$$f(2) = f(1 + 1) = f(1) + f(1) = c + c = 2c,$$

$$f(3) = f(2 + 1) = f(2) + f(1) = 2c + c = 3c,$$

and in general, for every natural number n,

$$f(n) = nc. \tag{2}$$

Further, $f(n + 0) = f(n) + f(0)$, and hence $f(0) = 0$. Consequently

$$f(-n) + f(n) = f(0) = 0,$$

so that (2) holds for every integer n. Moreover, analogous considerations for every integer $m > 0$ show that

$$f(n) = f\left(m \cdot \frac{n}{m}\right) = m \cdot f\left(\frac{n}{m}\right),$$

and hence

$$f\left(\frac{n}{m}\right) = \frac{n}{m} \cdot c,$$

so that

$$f(r) = rc$$

for every rational r. For a continuous function, however, this implies that $f(x) = cx$.

The question remained open, whether the functional equation (1) had other solutions besides those just found, if discontinuous functions were allowed. Hamel, now, has shown with the help of the well-ordering theorem, that there are then indeed still further solutions. The proof runs as follows:

In the domain of real numbers, a set $\mathfrak{B}(b)$ is called a basis if it has the following two properties:

a) For any finite number of basis elements $b_1, b_2, \cdots, b_m$, and arbitrary rational numbers $r_1, r_2, \cdots, r_m$ not all zero, we have never

$$\sum_{k=1}^{m} r_k b_k = 0.$$

b) For every number $z \neq 0$ there are among the basis elements a finite number of basis numbers $b_1, \cdots, b_m$, and, corresponding to these, rational numbers $r_1, \cdots, r_m$ all different from zero, such that

$$z = \sum_{k=1}^{m} r_k b_k .$$

First we have to show that such a basis exists. This is accomplished by building up the basis as a well-ordered set by means of transfinite induction. The starting-point is the well-ordering

$$\mathfrak{R} = \{A, B, C, \cdots\}$$

of the set of all real numbers. The number A shall belong to the basis if, and only if, $A \neq 0$. Suppose that it has already been decided for all the numbers of the segment $\mathfrak{R}_X$, which numbers shall belong to the basis and which shall not. Then X shall belong to the basis if, and only if, no equation of the kind described under a) subsists between X and any finite number of basis numbers already determined. In this way a set $\mathfrak{B}$ is defined by transfinite induction. $\mathfrak{B}$, as a subset of the well-ordered set $\mathfrak{R}$, is well-ordered by $\mathfrak{R}$.

This set $\mathfrak{B}$ is a basis. For suppose that an equation of the sort described under a) held for some finite number of basis numbers. Then one of these finitely many basis numbers would be the last to have a nonzero coefficient, and could then not be admitted into the basis, contrary to our assumption. Condition a) is thus fulfilled. That b) too is fulfilled is seen as follows: A given number $z \neq 0$ is either a basis number b or not. In the first case, b) is fulfilled trivially by the equation $z = b$. In the second case, an equation of the kind described in a) subsists between z and the basis numbers preceding it in $\mathfrak{R}$. On account of property a) already established for the basis numbers, z has a nonzero coefficient in this equation. The equation can therefore be solved for z, and if we leave out the terms with zero coefficients, assertion b) is obtained.

The representation which exists, according to b), for every

number $z \neq 0$, is uniquely determined by z. For if there were two distinct representations, subtraction of the two equations would result in a contradiction of a).

Now we define the function $f(x)$ first only for the basis numbers, and this is done in an arbitrary manner. Further, we let $f(0) = 0$. If $x \neq 0$ is an arbitrary real number, it can be represented in terms of a finite number of basis numbers with rational coefficients, in the form

$$x = \sum_k r_k b_k .$$

Here we shall also admit terms whose coefficients are zero; the uniqueness of the representation is then obviously still valid. Now we put

$$f(x) = \sum r_k f(b_k).$$

Then, if

$$y = \sum s_k b_k$$

is a representation of y in terms of basis elements,

$$x + y = \sum (r_k + s_k) b_k$$

is such a representation of $x + y$, and, as a matter of fact, due to the uniqueness, precisely *the* representation. Consequently,

$$f(x + y) = \sum (r_k + s_k) f(b_k) = \sum r_k f(b_k) + \sum s_k f(b_k)$$

$$= f(x) + f(y).$$

The function f is thus a solution of the functional equation. It is a discontinuous solution if we put, say, $f(b_1) = 0$, $f(b_2) = 1$ for two basis elements b_1, b_2; for if f were a continuous solution, the argument presented at the beginning shows that we should have to have $f(b_1) : f(b_2) = b_1 : b_2$, which is not the case.

Further relevant papers are cited, e. g., by Kamke, Jahresbericht der Deutschen Mathematiker-Vereinigung, vol. 36 (1927), pp. 145–156.

13. *The Well-ordering of Cardinal Numbers*

With the well-ordering theorem we immediately obtain from Theorem 3 on p. 88:

THEOREM 1. *If* $\mathfrak{m}$, $\mathfrak{n}$ *are two arbitrary cardinal numbers, precisely one of the three relations*

$$\mathfrak{m} < \mathfrak{n}, \quad \mathfrak{m} = \mathfrak{n}, \quad \mathfrak{m} > \mathfrak{n}$$

holds; i. e., any two cardinal numbers are comparable with each other.

Since the transitive law holds for the relations of magnitude of cardinal numbers, every set of cardinal numbers can be ordered according to increasing magnitude of its elements. This ordered set, however, will prove to be actually well-ordered. For this and similar considerations concerning cardinal numbers, it is practical to represent the cardinal numbers by means of ordinal numbers. According to the well-ordering theorem, every cardinal number can be represented by a well-ordered set; to put it briefly: every cardinal number $\mathfrak{m}$ can be represented by an ordinal number μ. The totality of ordinal numbers μ which can represent a cardinal number $\mathfrak{m}$ is designated as the number class[3] $\mathfrak{Z}(\mathfrak{m})$. This number class $\mathfrak{Z}(\mathfrak{m})$ contains, as does every class of ordinal numbers, a smallest ordinal number, which is called the initial number of $\mathfrak{Z}(\mathfrak{m})$, or the initial number belonging to $\mathfrak{m}$.

THEOREM 2. *Every transfinite initial number is a limit number.*

Proof: If some transfinite initial number were not a limit number, it would be immediately preceded by an ordinal number ν, and the initial number would be $\nu + 1$. But then ν and the initial number $\nu + 1$ would have the same cardinal number, i. e., $\nu + 1$ would not be the smallest ordinal number in its number class.

It is now possible to carry over §5, with the exception of

[3]Here, then, the $\mathfrak{m}$ in parentheses by no means represents the elements of $\mathfrak{Z}$.

Theorem 4, to cardinal numbers. In particular, the analogue of Theorems 1 and 3 reads as follows:

THEOREM 3. *Every set $\mathfrak{K}$ of cardinal numbers $\mathfrak{m}$, ordered according to increasing magnitude, is well-ordered. There exists a cardinal number which is greater than every cardinal number $\mathfrak{m}$ in $\mathfrak{K}$, and, in fact, there is a definite next larger cardinal number $\mathfrak{n}$.*

Proof: For the first part of the assertion we have to show that every nonempty subset $\mathfrak{K}_1$ of $\mathfrak{K}$ contains a smallest cardinal number. This is immediately clear, however, if every cardinal number is represented by the initial number belonging to it. Among these initial numbers then there is, as in every nonempty set of ordinal numbers, a smallest one, and the corresponding cardinal number is the smallest cardinal number in $\mathfrak{K}_1$.

That there exists a cardinal number which is greater than every cardinal number in $\mathfrak{K}$, was proved already on p. 35. Here we have merely to show, therefore, that there is a smallest cardinal number of this kind. Let $\overline{\mathfrak{m}}$ be a cardinal number which is greater than every $\mathfrak{m}$ in $\mathfrak{K}$. If $\overline{\mathfrak{m}}$ is not already itself the smallest cardinal number of this sort, form the set of cardinal numbers which are smaller than $\overline{\mathfrak{m}}$ and at the same time greater than every $\mathfrak{m}$ in $\mathfrak{K}$. This set contains, as does every nonempty set of cardinal numbers, a smallest element, and this proves the second part of the assertion.

If the number $\mathfrak{n}$ has no immediate predecessor, $\mathfrak{n}$ is again called a limit number and we write also

$$\mathfrak{n} = \lim_{\mathfrak{m}<\mathfrak{n}} \mathfrak{m}.$$

If the set of all transfinite cardinal numbers which are less than a given cardinal number $\mathfrak{n}$, is ordered according to increasing magnitude, then we denote the smallest (transfinite) cardinal number by $\aleph_0$,[4] the next, by $\aleph_1$, etc. In general, every cardinal number receives as index the ordinal number of the

[4] $\aleph$ = aleph, first letter of the Hebrew alphabet.

set of cardinal numbers which precede the cardinal number in question. In particular, therefore, we have invariably

$$\aleph_\mu < \aleph_\nu \quad \text{for} \quad \mu < \nu.$$

The initial number which belongs to $\aleph_\mu$ is denoted by ω_μ.

The smallest transfinite cardinal number was denoted before by $\mathfrak{a}$, and hence $\aleph_0 = \mathfrak{a}$. The power of the continuum is also denoted by $\aleph$ without an index. Certainly, then, $\aleph \geq \aleph_1$. The question whether the equality sign holds here is precisely the continuum problem. Further, ω_0 is the smallest transfinite limit number, and is therefore the same as the ω introduced earlier.

14. *Further Rules of Operation for Cardinal Numbers. Order Type of Number Classes*

Already in ch. II we proved the formulas $\mathfrak{a} \cdot \mathfrak{a} = \mathfrak{a}$ and $\mathfrak{a} \cdot \mathfrak{c} = \mathfrak{c}$. These are special cases of the following general rule:

$$\aleph_0 \cdot \aleph_\mu = \aleph_\mu. \tag{1}$$

Proof: The initial number which belongs to $\aleph_\mu$ is a limit number, and therefore, by p. 96, is divisible by ω; i. e., it can be represented in the form $\omega_\mu = \omega \cdot \nu$. If we set $|\nu| = \mathfrak{n}$, then accordingly

$$\aleph_\mu = \aleph_0 \, \mathfrak{n},$$

and hence indeed

$$\aleph_0 \aleph_\mu = \aleph_0 \aleph_0 \mathfrak{n} = \aleph_0 \mathfrak{n} = \aleph_\mu.$$

Likewise we already proved in ch. II that $\mathfrak{a} \cdot \mathfrak{a} = \mathfrak{a}$ and $\mathfrak{c} \cdot \mathfrak{c} = \mathfrak{c}$. These are special cases of the following general rule:

$$\aleph_\mu^2 = \aleph_\mu \qquad \text{(G. Hessenberg).} \tag{2}$$

Proof: The set of ordinal numbers $\xi < \omega_\mu$ has order type ω_μ, according to p. 86, Theorem 1, and is therefore a set of power $\aleph_\mu$. Consequently, the cardinal number $\aleph_\mu^2$ can be represented by the set of pairs (ξ, η) of all ordinal numbers $\xi < \omega_\mu$, $\eta < \omega_\mu$. We shall now show that, on the other hand, this set has at most the cardinal number $\aleph_\mu$. For this purpose we form the natural polynomial (p. 109)

$$\zeta = \sigma(\xi, \eta) = {}^*\sum_\nu \omega^\nu(x_\nu + y_\nu)$$

for $\xi < \omega_\mu$, $\eta < \omega_\mu$. For such a pair of numbers, let

$$\sigma(\xi, \eta) = \omega^\alpha(x + y) + \cdots,$$

starting with the highest term. Since only a finite number of nonzero terms appear, it follows that

$$(\alpha) \qquad \sigma(\xi, \eta) < \omega^\alpha(x + y + 1).$$

Since ω^α has a coefficient $\neq 0$ in at least one of the representations

$$\xi = {}^*\sum \omega^\nu x_\nu, \qquad \eta = {}^*\sum \omega^\nu y_\nu,$$

and since $\xi < \omega_\mu$, $\eta < \omega_\mu$, we have also $\omega^\alpha < \omega_\mu$. Accordingly, since ω_μ is an initial number, $|\omega^\alpha| < |\omega_\mu|$, and hence, since $x + y + 1$ is a finite number,

$$|\omega^\alpha(x + y + 1)| < |\omega_\mu|.$$

Consequently, from (α) follows also

$$\zeta = \sigma(\xi, \eta) < \omega_\mu.$$

We therefore certainly obtain all number pairs ξ, η by solving the equation $\sigma(\xi, \eta) = \zeta$ for ξ, η, for every $\zeta < \omega_\mu$. But for every ζ this equation has only a finite number of solutions (p. 110, Theorem 2). Therefore the cardinal number of the set of all solutions for all $\zeta < \omega_\mu$ is at most $\aleph_0 \cdot \aleph_\mu = \aleph_\mu$. Hence, $\aleph_\mu^2 \leq \aleph_\mu$, from which the assertion follows.

From these two fundamental rules we now obtain:

(3) For $\aleph_\mu \leq \aleph_\nu$ we have $\aleph_\mu + \aleph_\nu = \aleph_\mu \cdot \aleph_\nu = \aleph_\nu$,

and hence, in particular, $\aleph_\nu + \aleph_\nu = \aleph_\nu$.

For, $\aleph_\nu \leq \aleph_\mu + \aleph_\nu \leq 2\aleph_\nu \leq \aleph_\nu$,

and $$\aleph_\nu \leq \aleph_\mu \cdot \aleph_\nu \leq \aleph_\nu^2 = \aleph_\nu.$$

(4) For $\aleph_\rho < \aleph_\sigma$ and $\aleph_\mu < \aleph_\nu$ we have

$$\aleph_\rho + \aleph_\mu < \aleph_\sigma + \aleph_\nu \quad \text{and} \quad \aleph_\rho \cdot \aleph_\mu < \aleph_\sigma \cdot \aleph_\nu .$$

For let $\aleph_\rho \leq \aleph_\mu$, say. Then, by (3),

$$\aleph_\rho + \aleph_\mu = \aleph_\rho \cdot \aleph_\mu = \aleph_\mu < \aleph_\nu .$$

(5) $\sum_{\rho \leq \mu} \aleph_\rho = \aleph_\mu$, and, if μ is a limit number, also

$$\sum_{\rho < \mu} \aleph_\rho = \aleph_\mu .$$

For invariably $\aleph_\mu \leq \sum_{\rho \leq \mu} \aleph_\rho \leq \aleph_\mu \cdot \aleph_\mu = \aleph_\mu$. If μ is a limit number, then, on the one hand,

$$\sum_{\rho < \mu} \aleph_\rho \leq \aleph_\mu ,$$

and, on the other hand, for every $\nu < \mu$, also $\nu + 1 < \mu$, so that $\sum_{\rho < \mu} \aleph_\rho \geq \aleph_{\nu+1} > \aleph_\nu$; i. e., the sum is greater than every cardinal number which is less than $\aleph_\mu$. These two results then yield the assertion in this case too.

(6) $$\aleph_\mu^n = \aleph_\mu \quad \text{for finite } n > 0.$$

This follows immediately from (2).

In this formula, it is not possible, in general, to replace the exponent n by a transfinite cardinal number, such as $\aleph_0$, e. g. It is true that $\aleph^{\aleph_0} = \aleph$, and, for every $\aleph_\nu = \aleph_\mu^{\aleph_0}$,

$$\aleph_\nu^{\aleph_0} = \aleph_\mu^{\aleph_0 \aleph_0} = \aleph_\mu^{\aleph_0} = \aleph_\nu .$$

But $\aleph_0^{\aleph_0} > \aleph_0$, and generally, for an enumerable fundamental sequence of cardinal numbers with the limit $\aleph_\nu$,

$$\aleph_\nu^{\aleph_0} > \aleph_\nu ,$$

as follows from p. 46 with the aid of (5).

(7) $$2^{\aleph_\nu} = \aleph_\mu^{\aleph_\nu} \quad \text{for} \quad \mu \leq \nu + 1.$$

For, according to p. 43, b) and the hypothesis,

$$2^{\aleph_\nu} \geq \aleph_{\nu+1} \geq \aleph_\mu ,$$

and hence

$$2^{\aleph_\nu} = 2^{\aleph_\nu \aleph_\nu} = (2^{\aleph_\nu})^{\aleph_\nu} \geq \aleph_\mu^{\aleph_\nu} \geq 2^{\aleph_\nu}.$$

(8) $$\aleph_1^{\aleph_0} = \aleph.$$

Because $2^{\aleph_0} > \aleph_0$, i. e., $2^{\aleph_0} \geq \aleph_1$, implies that

$$\aleph = 2^{\aleph_0} \leq \aleph_1^{\aleph_0} \leq (2^{\aleph_0})^{\aleph_0} = 2^{\aleph_0} = \aleph.$$

For the proofs of two additional aleph relations, we need the following

Lemma. Let $\mathfrak{K}(\rho_k)$ *be a well-ordered set of ordinal numbers* ρ_k, $|\mathfrak{K}| \leq \aleph_{\mu-1}$, *and every* $|\rho_k| \leq \aleph_{\mu-1}$. *Then there exists an ordinal number* $\sigma < \omega_\mu$ *such that every* $\rho_k < \sigma$.

Proof: Put $\sigma = \sum_{k \in \mathfrak{K}} (\rho_k + 1)$. Then $\sigma \geq$ every $\rho_k + 1$, and hence $\sigma >$ every ρ_k, and $|\sigma| \leq \sum_{k \in \mathfrak{K}} |\rho_k + 1| \leq \aleph_{\mu-1} \aleph_{\mu-1} = \aleph_{\mu-1}$. Consequently, since ω_μ is the initial number of $\mathfrak{Z}(\aleph_\mu)$, also $\sigma < \omega_\mu$.

(9) $\aleph_\mu^{\aleph_\nu} = \aleph_\mu \aleph_{\mu-1}^{\aleph_\nu}$, if μ is not a limit number (Hausdorff).

Proof: For $\mu \leq \nu$ it follows from (7) that

$$2^{\aleph_\nu} = \aleph_\mu^{\aleph_\nu} = \aleph_{\mu-1}^{\aleph_\nu}.$$

Multiplication by $\aleph_\mu$ yields

$$\aleph_\mu^{\aleph_\nu} = \aleph_\mu \aleph_\mu^{\aleph_\nu} = \aleph_\mu \aleph_{\mu-1}^{\aleph_\nu}.$$

For $\mu \leq \nu$, this already proves the assertion. Let us assume, then, that $\mu > \nu$, and first prove that

(β) $$\aleph_\mu^{\aleph_\nu} \leq \sum_{\sigma < \omega_\mu} |\sigma|^{\aleph_\nu}$$

if μ is not a limit number. Let $\mathfrak{M}$ be the set of ordinal numbers $\rho < \omega_\mu$, and let $\mathfrak{N}$ be the set of ordinal numbers $< \omega_\nu$; the elements of $\mathfrak{N}$ will also be denoted by n. For these sets, since μ is not a limit number, we have

(γ) $$|\mathfrak{M}| = |\omega_\mu| = \aleph_\mu, \quad |\mathfrak{N}| = |\omega_\nu| = \aleph_\nu \leq \aleph_{\mu-1},$$

and

$(\delta)\qquad |\rho| < |\omega_\mu|,\quad$ so that $\quad |\rho| \leq \aleph_{\mu-1}\quad$ for $\quad \rho \in \mathfrak{M}.$

The left-hand side of (β) can be represented by the covering set (cf. p. 42) $\mathfrak{N} \mid \mathfrak{M}$. Every element of $\mathfrak{N} \mid \mathfrak{M}$ is a set $\mathfrak{N}((n, \rho))$. For a fixed element of this sort, it follows from the lemma, because of (γ) and (δ), that all ρ's of this element belong to a set $\mathfrak{W}_\sigma$ with $\sigma < \omega_\mu$. Therefore the element $\mathfrak{N}((n, \rho))$ also appears in the covering set $\mathfrak{N} \mid \mathfrak{W}_\sigma$ for a certain $\sigma < \omega_\mu$. The covering set $\mathfrak{N} \mid \mathfrak{M}$ is thus contained in the totality of all covering sets $\mathfrak{N} \mid \mathfrak{W}_\sigma$ arising from all $\sigma < \omega_\mu$. This fact implies (β).

From (β) now follows, since, in the sum, every $\sigma < \omega_\mu$ and hence $|\sigma| \leq \aleph_{\mu-1}$,

$$\aleph_\mu^{\aleph_\nu} \leq |\omega_\mu| \cdot \aleph_{\mu-1}^{\aleph_\nu} = \aleph_\mu \aleph_{\mu-1}^{\aleph_\nu} \leq \aleph_\mu \aleph_\mu^{\aleph_\nu} = \aleph_\mu^{\aleph_\nu},$$

which proves (9).

(10) For $\mu \leq \nu + n$, where n denotes an integer ≥ 0, we have

$$\aleph_\mu^{\aleph_\nu} = 2^{\aleph_\nu} \cdot \aleph_\mu \qquad \text{(F. Bernstein).}$$

Proof: For $\mu \leq \nu$, the assertion follows from (7) after multiplication by $\aleph_\mu$. The other cases follow by mathematical (finite) induction. For if the assertion is already true for n, then, for $\mu = \nu + n + 1$ in (β),

$$|\sigma|^{\aleph_\nu} = 2^{\aleph_\nu} |\sigma|,$$

and hence it follows from (β) that

$$\aleph_\mu^{\aleph_\nu} \leq 2^{\aleph_\nu} \aleph_\mu^2 = 2^{\aleph_\nu} \aleph_\mu \leq \aleph_\mu^{\aleph_\nu} \aleph_\mu = \aleph_\mu^{\aleph_\nu}.$$

The number class $\mathfrak{Z}(\aleph_\mu)$ was defined on p. 119 as the set of those ordinal numbers which have power $\aleph_\mu$. The number class is thus a set of ordinal numbers, and we can therefore inquire as to its order type and its power. The answer is furnished by the following

THEOREM. *The number class $\mathfrak{Z}(\aleph_\mu)$, ordered according to in-*

creasing magnitude of its elements, has order type $\omega_{\mu+1}$ *, and therefore has cardinal number* $\aleph_{\mu+1}$.

Proof: The elements of $\mathfrak{Z}(\aleph_\mu)$ are the ordinal numbers ν with

$$\omega_\mu \leq \nu < \omega_{\mu+1} .$$

Hence, in the sense of addition of well-ordered sets,

$$\mathfrak{W}_{\omega_\mu} + \mathfrak{Z}(\aleph_\mu) = \mathfrak{W}_{\omega_{\mu+1}} .$$

If ζ is the ordinal number of $\mathfrak{Z}(\aleph_\mu)$, this equation implies

$$(\epsilon) \qquad \omega_\mu + \zeta = \omega_{\mu+1} \quad \text{and} \quad \aleph_\mu + |\zeta| = \aleph_{\mu+1} .$$

From this follows $|\zeta| \leq \aleph_{\mu+1}$. But, on account of (4), we cannot have $|\zeta| < \aleph_{\mu+1}$. Therefore $|\zeta| = \aleph_{\mu+1}$, so that $\zeta \geq \omega_{\mu+1}$, and hence, because of (ϵ), we have indeed $\zeta = \omega_{\mu+1}$.

Cantor called the set of all finite ordinal numbers the first number-class, and the set of ordinal numbers with cardinal number $\aleph_0$, the second number-class. We just showed that this number class has power $\aleph_1$. The continuum problem can accordingly be formulated also as follows: Does the second number-class have power $\aleph$?

15. Ordinal Numbers and Sets of Points

Cantor made use of ordinal numbers in order to define, for sets of points, derived sets of arbitrarily high order. We shall restrict ourselves here to point sets which lie in the plane, because the matters can be visualized best in the plane. The following considerations, however, can be carried over immediately to space.

By an ϵ-neighborhood or briefly a neighborhood of a point P_0 we mean the set of all points P whose distance $\overline{PP_0}$ from P_0 is less than ϵ; in other words, the interior of the circle with center P_0 and radius ϵ. If a point set $\mathfrak{M}$ is given, a point P, irrespective of whether it belongs to $\mathfrak{M}$ or not, is called a limit point of $\mathfrak{M}$, if every neighborhood of P contains infinitely many points of $\mathfrak{M}$, or, what amounts to the same thing, if every neighborhood of P contains at least one point which belongs to $\mathfrak{M}$ and differs from P. For example, if $\mathfrak{M}$ is the set of all points, both of whose coordinates are rational numbers, then every point of the plane is a limit point of $\mathfrak{M}$. On the other hand, the set of points, both of whose coordinates are integers, has no limit points. Likewise, no finite set has a limit point.

By the first derived set $\mathfrak{M}'$ of a set $\mathfrak{M}$ we mean the set of limit points of $\mathfrak{M}$, provided that such points exist. If $\mathfrak{M}$ has no limit point, we put $\mathfrak{M}' = 0$. The second derived set $\mathfrak{M}''$ of $\mathfrak{M}$ shall be the derived set of $\mathfrak{M}'$, so that $\mathfrak{M}'' = (\mathfrak{M}')'$. By continuing this definition, we can thus ascend to derived sets $\mathfrak{M}^{(n)}$ of arbitrary finite order. By transfinite induction, however, we can ascend also to derived sets $\mathfrak{M}^{(\nu)}$ whose order ν is an arbitrary ordinal number. If the derived sets are already defined for all $\mu < \nu$, we put

$$\mathfrak{M}^{(\nu)} = \begin{cases} (\mathfrak{M}^{(\nu-1)})', \text{ if } \nu \text{ is not a limit number;} \\ \text{the intersection of all } \mathfrak{M}^{(\mu)}\text{'s for all } \mu < \nu, \text{ if } \nu \text{ is a limit number.} \end{cases}$$

Accordingly, we have, e. g.,

$$\mathfrak{M}^{(\omega)} = \mathfrak{M}' \cdot \mathfrak{M}'' \cdot \mathfrak{M}''' \cdots, \quad \mathfrak{M}^{(\omega+1)} = (\mathfrak{M}^{(\omega)})'.$$

Actually, however, there is no point in ascending to derived sets of arbitrarily high order. For, as Cantor proved, the process automatically terminates with an ordinal number of the second number-class $\mathfrak{Z}(\aleph_0)$ in the sense that all subsequent derived sets are identical with one another. The proof of this theorem is the main goal of the following discussion. For the proof we have to introduce a few more concepts from the theory of point sets.

First let us sharpen the concept of limit point: If a point set $\mathfrak{M}$ is given, a point P, irrespective of whether it belongs to $\mathfrak{M}$ or not, is called a condensation point of $\mathfrak{M}$, if every neighborhood of P contains a nonenumerable number of points of $\mathfrak{M}$.

If, e. g., $\mathfrak{M}$ is the set of points on the number axis which correspond to the irrational numbers, then every point of the number axis is a condensation point of $\mathfrak{M}$.

Regarding the existence of limit and condensation points, there are the following two theorems:

THEOREM 1 (Bolzano-Weierstrass). *Every bounded*[5] *infinite point set $\mathfrak{M}$ has at least one limit point.*

Proof: Since $\mathfrak{M}$ is bounded, $\mathfrak{M}$ lies completely within some closed square $\mathfrak{Q}$. Subdivide the latter into four congruent squares, and consider each of these as closed. Since $\mathfrak{M}$ is infinite, at least one of these four subsquares also contains infinitely many points of $\mathfrak{M}$; let $\mathfrak{Q}_1$ denote such a square. Now we again quarter $\mathfrak{Q}_1$, and denote by $\mathfrak{Q}_2$ one of the quarters containing infinitely many points of $\mathfrak{M}$; etc. We thus obtain a sequence of nested squares whose sides tend to the limit 0. This nest of squares therefore determines a point P, and this is a limit point of $\mathfrak{M}$, because every neighborhood of P certainly contains one of the squares $\mathfrak{Q}_n$ completely, and the latter contains infinitely many points of $\mathfrak{M}$.

Note that the proof asserts nothing as to whether P belongs to $\mathfrak{M}$ or not.

THEOREM 2. *Every nonenumerable point set $\mathfrak{M}$ has at least one condensation point.*

Proof: First, let $\mathfrak{M}$ be bounded. Then the assertion follows by replacing the words "limit point" and "infinitely many" in the preceding proof, by "condensation point" and "nonenumerably many". If $\mathfrak{M}$ is not bounded, we divide the plane up into the squares

$$a \leq x < a + 1, \quad b \leq y < b + 1,$$

where a and b run through all the integers. At least one of these squares contains nonenumerably many points of $\mathfrak{M}$, because otherwise $\mathfrak{M}$, as the union of the subsets of $\mathfrak{M}$ lying in the enumerably many squares, would be at most enumerable,

[5]A point set is said to be bounded, if it lies entirely within some square.

contrary to hypothesis. But then the part of the theorem already proved again yields at least one condensation point.

The concepts "limit point" and "condensation point" are closely related to two other concepts. Between a point set $\mathfrak{M}$ and its first derived set $\mathfrak{M}'$, the following relations, among others, are conceivable:

$$\text{a) } \mathfrak{M}' \subseteq \mathfrak{M} \quad \text{or} \quad \text{b) } \mathfrak{M} \subseteq \mathfrak{M}'.$$

Both cases actually occur too. On the number axis, the set $\{0, \frac{1}{2}, \frac{1}{3}, \frac{1}{4}, \cdots\}$ furnishes an example of case a), and the set of rational numbers is an example of the second case. We are therefore entitled to make the following definition:

A point set $\mathfrak{M}$ is said to be closed, if $\mathfrak{M}' \subseteq \mathfrak{M}$. It is said to be dense in itself, if $\mathfrak{M} \subseteq \mathfrak{M}'$ and $\mathfrak{M} \neq 0$. It is said to be perfect, if $\mathfrak{M}' = \mathfrak{M}$ and $\mathfrak{M} \neq 0$.

Every closed interval is a closed, and actually a perfect, set. Every open interval is dense in itself, but not closed, and therefore also not perfect. Every finite set is closed, because its derived set is the empty set, and the latter is a subset of every set. Every finite set, however, is not dense in itself, and therefore also not perfect.

THEOREM 3. *The intersection $\mathfrak{D}$ of arbitrarily many closed sets $\mathfrak{M}$ is closed.*

Proof: If $\mathfrak{D}' = 0$, the assertion is trivially correct. Suppose, then, that D is a limit point of $\mathfrak{D}$. Then every neighborhood of D contains infinitely many points of $\mathfrak{D}$, and hence also infinitely many points of every $\mathfrak{M}$. This means that D is a limit point of every $\mathfrak{M}$, and hence, since the $\mathfrak{M}$'s are closed, a point of every $\mathfrak{M}$. D therefore is also a point of $\mathfrak{D}$; i. e., $\mathfrak{D}$ is closed.

THEOREM 4. *Every derived set $\mathfrak{M}^{(\nu)}$ of a point set $\mathfrak{M}$ is closed.*

Proof: We first prove the assertion for the first derived set, where we may assume that the next derived set $\mathfrak{M}'' \neq 0$. Now let P'' be any point of $\mathfrak{M}''$. We have to show that P'' belongs to $\mathfrak{M}'$, i. e., that P'' is a limit point of $\mathfrak{M}$. In fact, given $\epsilon > 0$, there exists at least one point P' of $\mathfrak{M}'$ in the $\frac{1}{2}\epsilon$-neighborhood of P'', because P'' is a limit point of $\mathfrak{M}'$. Since P' is a limit point of $\mathfrak{M}$, there are infinitely many points of $\mathfrak{M}$ in the $\frac{1}{2}\epsilon$-neighborhood of P', and hence *a fortiori*, in the ϵ-neighborhood of P''. From the fact that the next derived set of every set is closed, however, the general validity of the theorem follows by transfinite induction. For we just proved the theorem for the first derived set. Suppose that it has already been proved for all $\mathfrak{M}^{(\mu)}$'s with $\mu < \nu$. If ν is not a limit number, then $\mathfrak{M}^{(\nu)}$, as the derived set of $\mathfrak{M}^{(\nu-1)}$, is closed, according to the part just proved. If, however, ν is a limit number, then $\mathfrak{M}^{(\nu)}$ is the intersection of all the preceding derived sets, i. e., the intersection of closed sets, and the assertion then follows from Theorem 3.

The derived set is the set of limit points. For condensation points, the analogue of Theorem 4 reads as follows:

THEOREM 5. *For every nonenumerable set $\mathfrak{M}$, the set of its condensation points is perfect.*

Proof: Let $\mathfrak{B}$ be the set of condensation points of $\mathfrak{M}$. We shall first prove that $\mathfrak{B}$ is closed, and in doing so we may assume that $\mathfrak{B}' \neq 0$. Every neighborhood of a point V' of $\mathfrak{B}'$ contains at least one point of $\mathfrak{B}$. This neighborhood, therefore, by the same argument as in the preceding proof, contains also nonenumerably many points of $\mathfrak{M}$. Consequently, V' belongs to $\mathfrak{B}$, i. e., $\mathfrak{B}$ is closed. On the other hand, $\mathfrak{B}$ is also dense in itself, i. e., every neighborhood of an arbitrary point V of $\mathfrak{B}$ contains at least one other point of $\mathfrak{B}$. For if this were not the case, there would exist an ϵ-neighborhood of V containing no other point of $\mathfrak{B}$. By successively halving ϵ, we construct a sequence of concentric circles about V, whose radii tend to the limit 0. These circles determine a sequence of circular rings l_1 , l_2 , $\cdots$. According to our assumption, no circular ring contains a point of $\mathfrak{B}$, i. e., every circular ring contains at most enumerably many points of $\mathfrak{M}$. Since the sum of the circular rings, however, gives the ϵ-neighborhood of V except for the point V itself, the ϵ-neighborhood of V would also contain only enumerably many points of $\mathfrak{M}$, i. e., V would not be a condensation point of $\mathfrak{M}$, contrary to our assumption.

The converses of the two theorems just proved are also true; i. e., it can be shown that every closed set of points can be regarded as the derived set of some set, and that every perfect set coincides with the set of its condensation points. We shall not, however, make any use of these facts. On the other hand, we do need

THEOREM 6. *Every perfect set of points has power* $\aleph$.

Proof: The set of all the points of the plane has power $\aleph$, and the power of the given perfect set $\mathfrak{M}$ is therefore at most $\aleph$. We have merely to prove, then, that $|\mathfrak{M}| \geq \aleph$. Since $\mathfrak{M}$ is a perfect set, it certainly contains infinitely many points. Let P_0 , P_1 be two points of $\mathfrak{M}$. Describe mutually external circles $\mathfrak{K}_0$, $\mathfrak{K}_1$ about these two points. Since $\mathfrak{M}$ is dense in itself, there are two points P_{00} , P_{01} in $\mathfrak{K}_0$ which belong to $\mathfrak{M}$. About each of these two points we again describe mutually external circles $\mathfrak{K}_{00}$, $\mathfrak{K}_{01}$, which lie within $\mathfrak{K}_0$ and have radii that are at most equal to half the radius of $\mathfrak{K}_0$. We do the same in $\mathfrak{K}_1$, and the circles obtained there are denoted by $\mathfrak{K}_{10}$, $\mathfrak{K}_{11}$. We can continue in this fashion. In every circle, about two points of $\mathfrak{M}$ we always describe circles which lie in the preceding circle, are mutually external, and have radii that are at most equal to half the radius of the preceding circle. The new circles receive as indices the index of the preceding circle with an added 0 or 1. We thus obtain, for every terminating dyadic fraction

$$d_n = 0 \cdot a_1 a_2 \cdots a_n , \quad \text{a circle} \quad \mathfrak{K}_{a_1 a_2 \cdots a_n} .$$

If

$$d_m = 0 \cdot a_1 \cdots a_n a_{n+1} \cdots a_m ,$$

then $\mathfrak{K}_{a_1 \cdots a_m}$ is contained in $\mathfrak{K}_{a_1 \cdots a_n}$, and the diameters of the circles $\mathfrak{K}_{a_1 \cdots a_n}$ tend to the limit 0 as $n \rightarrow \infty$.

With every infinite dyadic fraction (which may, from a certain point on, contain nothing but zeros) $0 \cdot a_1 a_2 a_3 \cdots$, there is thus associated a sequence of nested circles which, because of the limit relation for their radii, determine precisely one point P belonging to all the circles of this sequence. Since every circle contains points of $\mathfrak{M}$, P is a limit point of $\mathfrak{M}$; and due to the fact that $\mathfrak{M}$ is closed, P is actually a point of $\mathfrak{M}$. Further, from the way in which the circles were constructed, it follows that two distinct dyadic fractions determine two distinct points of $\mathfrak{M}$. The set $\mathfrak{M}$ therefore contains a subset which is equivalent to the set of all dyadic fractions $0 \cdot a_1 a_2 a_3 \cdots$. Since the set of these dyadic fractions has power $2^{\aleph_0} = \aleph$, $|\mathfrak{M}| \geq \aleph$, and the theorem is proved.

After these preparations, we now turn to the proposition formulated above on p. 127, and we prove a general theorem from which the aforementioned proposition easily follows.

THEOREM 7 (R. Baire). *For every ordinal number α of the first or second number class, let a closed set $\mathfrak{M}_\alpha$ be defined, and let*

$$\mathfrak{M}_\beta \subseteq \mathfrak{M}_\alpha \quad \textit{for} \quad \alpha < \beta.$$

Then there exists an ordinal number γ of the first or second number class, such that

$$\mathfrak{M}_\alpha = \mathfrak{M}_\gamma \quad \textit{for every} \quad \alpha \geq \gamma.$$ [6]

Proof: Let α be a number for which a $\delta > \alpha$ exists such that $\mathfrak{M}_\delta \subset \mathfrak{M}_\alpha$, and in fact let δ be the smallest number of this kind. Among the defined $\mathfrak{M}$'s which satisfy this relation for this δ and which are equal to $\mathfrak{M}_\alpha$, there is one whose index is smallest, and we shall assume that this $\mathfrak{M}$ is already $\mathfrak{M}_\alpha$ to begin with. Then $\mathfrak{M}_\alpha$ contains a point P which does not belong to $\mathfrak{M}_\delta$, and hence, since the sets are closed, $\mathfrak{M}_\alpha$ contains a neighborhood of P which does not belong to $\mathfrak{M}_\delta$. There is therefore a circle $\mathfrak{k}_\alpha$ which contains P but no point of $\mathfrak{M}_\delta$, and which has a rational radius as well as a center whose coordinates are both rational numbers.

Let $\alpha < \beta$ be two ordinal numbers of the first or second number class, for which $\mathfrak{k}_\alpha$ and $\mathfrak{k}_\beta$ exist. Then $\mathfrak{k}_\alpha \neq \mathfrak{k}_\beta$. For there is then a smallest number $\delta > \alpha$ for which $\mathfrak{M}_\delta \subset \mathfrak{M}_\alpha$, and $\delta \leq \beta$. Since $\mathfrak{k}_\alpha$ contains no point of $\mathfrak{M}_\delta$, and since $\mathfrak{M}_\beta \subseteq \mathfrak{M}_\delta$, $\mathfrak{k}_\alpha$ contains no point of $\mathfrak{M}_\beta$. But $\mathfrak{k}_\beta$ contains at least one point of $\mathfrak{M}_\beta$, and therefore $\mathfrak{k}_\alpha \neq \mathfrak{k}_\beta$.

---

[6]For further theorems of a similar nature, see Hausdorff [2], p. 170.

According to the last example on p. 9, there are only an enumerable number of circles with rational radii and centers whose coordinates are rational numbers. Consequently, there exist also merely an enumerable number of the above $\mathfrak{k}_\alpha$'s; i. e., the indices of the $\mathfrak{k}_\alpha$'s form an enumerable set of ordinal numbers of the first or second number class. By p. 91, Theorem 4 there is then an ordinal number γ of the first or second number class, such that $\alpha < \gamma$ for each of these α's. Thus, for every $\alpha \geq \gamma$, no $\mathfrak{k}_\alpha$ exists; i. e., for these α's, $\mathfrak{M}_\alpha = \mathfrak{M}_\gamma$.

From Baire's theorem now follows

THEOREM 8 (Cantor-Bendixson). *Let $\mathfrak{M}$ be a closed set. Then there exists a number γ of the first or second number class, such that $\mathfrak{M}^{(\gamma)}$ is either empty or perfect. Further, $\mathfrak{M}^{(\alpha)} = \mathfrak{M}^{(\gamma)}$ for $\alpha \geq \gamma$, and $\mathfrak{M}^{(\gamma)} = 0$ or $\supset 0$ according as $\mathfrak{M}$ is enumerable or not. In every case, $\mathfrak{M}$ can be represented in the form $\mathfrak{M} = \mathfrak{M}^{(\gamma)} + \mathfrak{R}$, where $\mathfrak{R}$ is at most enumerable and $\mathfrak{R} \cdot \mathfrak{M}^{(\gamma)} = 0$.*

Proof: Since, according to Theorem 4, every derived set is closed, it follows from the definition of derived set, that the derived sets $\mathfrak{M}^{(\alpha)}$ satisfy the assumptions made concerning the $\mathfrak{M}_\alpha$'s in the preceding theorem. Hence, there exists a number γ of the first or second number class, such that $\mathfrak{M}^{(\alpha)} = \mathfrak{M}^{(\gamma)}$ for $\alpha \geq \gamma$. $\mathfrak{M}^{(\gamma)}$, as a derived set, is closed; but it is also either dense in itself or empty, because otherwise $\mathfrak{M}^{(\gamma)}$ would contain a point which was not a limit point of $\mathfrak{M}^{(\gamma)}$, and this point would therefore disappear if we went over to the next derived set, so that we should have $\mathfrak{M}^{(\gamma+1)} \neq \mathfrak{M}^{(\gamma)}$. Consequently, $\mathfrak{M}^{(\gamma)}$ is either empty or perfect. If $\mathfrak{M}^{(\gamma)} \supset 0$, and hence is perfect, then, since $\mathfrak{M}^{(\gamma)} \subseteq \mathfrak{M}$, $\mathfrak{M}$ contains a perfect subset, and therefore, by Theorem 6, $\mathfrak{M}$ is nonenumerable. Conversely, if $\mathfrak{M}$ is nonenumerable, then, according to Theorems 2 and 5, $\mathfrak{M}$ contains a perfect subset which then belongs also to all the derived sets, so that $\mathfrak{M}^{(\gamma)} \supset 0$.

Further, $\mathfrak{R}$ is at most enumerable. For let $\mathfrak{R}_n$ be the set of those points of $\mathfrak{R}$ which have a distance $\geq 1/n$ from every point of $\mathfrak{M}^{(\gamma)}$. Then $\mathfrak{R}_n$ is a closed set, and is also enumerable. For if $\mathfrak{R}_n$ were not enumerable, application of that part of our theorem which has already been proved would show that $\mathfrak{N}_n$ contained a perfect set, which $\mathfrak{M}^{(\gamma)}$ would then have to contain in addition. Since every single point of the set $\mathfrak{N}$ has a positive distance from $\mathfrak{M}^{(\gamma)}$ (this follows from the fact that $\mathfrak{M}^{(\gamma)}$ is closed), $\mathfrak{N}$ is the sum of all the $\mathfrak{N}_n$'s for $n = 1, 2, 3, \cdots$. $\mathfrak{N}$ is thus the sum of enumerably many sets, each of which is at most enumerable, so that $\mathfrak{N}$ itself is at most enumerable.

Finally, $\mathfrak{N} \cdot \mathfrak{M}^{(\gamma)} = 0$, and hence, since $\mathfrak{N} \subseteq \mathfrak{M}$ and therefore $\mathfrak{N}^{(\gamma)} \subseteq \mathfrak{M}^{(\gamma)}$, *a fortiori* also $\mathfrak{N} \cdot \mathfrak{N}^{(\gamma)} = 0$.

For Cantor, the motive for the investigations described here was the continuum problem, which, in the more general form of whether there always exists another cardinal number between $\aleph_\nu$ and $2^{\aleph_\nu}$, is the central problem of the theory of sets. The cardinal number $\aleph$ can be represented by the points of the interval $\langle 0, 1\rangle$. If $\aleph_1 < \aleph$, the interval $\langle 0, 1\rangle$ must contain a point set of cardinal number $\aleph_1$. The Cantor-Bendixson theorem shows now that such a set, if it exists at all, i. e., if $\aleph_1 < \aleph$, at all events cannot be a closed set. For from

$$\mathfrak{M} = \mathfrak{M}^{(\gamma)} + \mathfrak{N}$$

it follows, since $\mathfrak{M}^{(\gamma)}$ is either empty or perfect, and hence, by Theorem 6, has either cardinal number 0 or $\aleph$, that every closed set has a cardinal number which is either $\leq \aleph_0$ or $= \aleph$.

Concluding Remarks

We have already occasionally pointed out several contradictions which arise in pursuing the concepts employed in the theory of sets. These contradictions were connected with the following concepts:

a) The set of all cardinal numbers (p. 36).

b) The set of all ordinal numbers (Burali-Forti's Paradox, p. 91).

c) The set of all even ordinal numbers (p. 96).

To these the following paradoxes can be added:

d) The set of all sets which do not contain themselves as elements (Zermelo-Russell).

For this set $\mathfrak{M}$, as for every set, only the following two cases are conceivable: either $\mathfrak{M}$ contains itself as an element, or it does not. The first case cannot occur, because otherwise $\mathfrak{M}$ would have an element, namely, $\mathfrak{M}$, which contained itself as an element. The second assumption, however, also leads to a contradiction. For in this case, $\mathfrak{M}$ would be a set which did not contain itself as an element, and for this very reason, $\mathfrak{M}$ would be contained in the totality of *all* such sets, i. e., in $\mathfrak{M}$.

e) The set of all sets.

For if $\mathfrak{M}$ is this set, it would have to contain the set of all its subsets. The set of all subsets of $\mathfrak{M}$, however, has a greater power than $\mathfrak{M}$.

Here we have aided ourselves by erecting danger signs before the paradoxical notions. All proofs have been carried out so as to avoid those ideas which have hitherto been recognized as self-contradictory.

Such a procedure is justifiable if the purpose is to acquaint someone for the first time with the results and methods of proof peculiar to the theory of sets. But a rigorous and final construction of set theory cannot be attained in this way. For we should have to fear the possible existence of more paradoxes (in fact, a recipe is given below for the construction of additional paradoxes).

In the paradoxes listed above, it strikes one that the word "all" appears in each of them. As a result, the set e), if it is meaningful, is a set which contains itself as an element. The same holds for the set d). For if $\mathfrak{M}$ is any set whose elements are *exclusively* sets which do not contain themselves as elements, then $\mathfrak{M}$ cannot be an element of itself. This argument has nothing to do yet with the word "all". Not until the second part of the argument given in connection with d) is a conclusion drawn from the occurrence of the word "all", this inference being that the set d) would have to contain itself as an element, which then leads to a contradiction.

These two samples lead one to conjecture that all sets which contain themselves as elements are, as a matter of course, self-contradictory concepts, and are therefore inadmissible. In fact, up to now no sets are known which contain themselves as elements and which everyone would regard without hesitation as meaningful sets. Also, Cantor's definition of a set as "a collection into a whole of definite, well-distinguished objects" must be interpreted to mean that something new is created by this act of collecting, so that a set can never equal one of its elements.

If we do not admit sets, then, which contain themselves as elements, paradoxes d) and e) fall to the ground, since the sets are inadmissible according to this criterion. The elimination of the remaining paradoxes by means of this criterion, however, is not possible in a satisfactory manner. One can say that cardinal numbers and ordinal numbers are nothing but sets, and that in order to arrive, e. g., at the set $\mathfrak{K}$ of all cardinal numbers, one would first have to form the set of all sets, which has already been recognized as meaningless, and then pick out a representative from every class of mutually equivalent sets, in order to obtain the elements of $\mathfrak{K}$. But one is by no means forced to take this path: In order to arrive at cardinal and ordinal numbers, we do not have to consider sets of an arbitrary sort, but, on the contrary, may restrict ourselves to sets with well-defined mathematical objects as elements. For the first few ordinal numbers, such representations by means of the set $\{0, 1, 2, \cdots\}$ are given on p. 86.

If, in cases a) to c), we order the set according to increasing magnitude of its elements, then the contradiction becomes immediately apparent if we ascribe an ordinal number to this fundamental sequence. There is, however, an analogue of this for ordinary sequences of real numbers in analysis. A monotonically increasing sequence of real numbers converges only if it is bounded, i. e., if its elements are less than a fixed real number. If one were to assume that every monotonically increasing sequence of real numbers converged to a real number, one would soon arrive at contradictions. It appears to be similar in the theory of sets. For not only do the antinomies a) to e) disappear when we admit as elements of sets only such sets, ordinal numbers, and cardinal numbers as are bounded above by a fixed cardinal number, but we see also that paradoxes always arise if we collect into a set any sets, cardinal numbers, or ordinal numbers which are not bounded above by a fixed cardinal number. Thus we have here a recipe for constructing additional paradoxes. We can form, e. g., the set of all ordinal numbers which are divisible by some fixed ordinal number α, or the set of all initial numbers, or the set of all sets which have a last element, etc. All these sets prove to be self-contradictory.

Whereas the given principle removes all hitherto known paradoxes of set theory,[1] this solution nevertheless cannot be regarded as satisfactory, because a mere argument by analogy, such as led us here to our principle, possesses no force of conviction, but can only give us hints as to the direction in which, perhaps, the solution of the problem is to be sought. One is reminded of the contradictions which the history of mathematics shows attended unscrupulous operation with infinite series and sequences, and which were subsequently removed by

[1]In the development of the theory of sets, some other paradoxes, to be sure, besides those mentioned above play a role. But these do not belong basically to set theory, but rather to general logic, and their resolution is therefore also to be expected only from this direction. For this reason they may be disregarded here.

clarifying the concept of number and the circumstances under which a sequence or series can represent a number. What is of foremost importance, then, is the production of the scales of cardinal and ordinal numbers in a rigorous fashion. If this is attempted by means of a constructive principle, it turns out (cf. Enzyklopädie, I_1, Heft 2, article 5, no. 27) that certain principles of construction lead to a very comprehensive domain $\mathfrak{W}_I$ of ordinal numbers (and hence also of cardinal numbers), but that, in accordance with our resolution above of the paradoxes, no ordinal number of the domain $\mathfrak{W}_I$ is the ordinal number of the set of *all* ordinal numbers of this domain. If we extend the construction principles in a suitable manner, we obtain an even more comprehensive domain $\mathfrak{W}_{II}$ of ordinal numbers, but once again none of the ordinal numbers defined thus far corresponds to the set of *all* ordinal numbers of this domain. This observation repeats itself on continuing such a constructive development. It corresponds exactly to the aforementioned fact that we encounter paradoxes if we consider boundless sets of cardinal or ordinal numbers.

For further critical remarks concerning the foundations of the theory of sets, the reader is referred to Enzyklopädie, I_1, Heft 2, article 5, nos. 11–16.

Bibliography

For more extensive studies of the whole field of the theory of sets, the reader is referred to:

Hausdorff, F., [1] *Grundzüge der Mengenlehre*, 1st edition (reprint), New York, 1949.

————, [2] *Mengenlehre*, 3d edition (reprint), New York, 1944.

Sierpiński, W., *Leçons sur les nombres transfinis*, Paris, 1928.

————, *Hypothèse du continu*, Warszawa-Lwów, 1934.

The reader who is more interested in the development of the principles should consult:

Fraenkel, A., [1] *Einleitung in die Mengenlehre*, 3d edition (reprint), New York, 1946.

————, [2] *Zehn Vorlesungen über die Grundlegung der Mengenlehre*, Leipzig and Berlin, 1927.

We mention, in addition, the following reference works:

Enzyklopädie der Mathematischen Wissenschaften, Band I_1, Heft 2, 2d edition, Leipzig and Berlin, 1939.

Pascal, E., *Repertorium der höheren Mathematik*, vol. 1, 2d edition, parts 1 and 3, Leipzig and Berlin, 1910, 1928.

Schoenflies, A., *Die Entwickelung der Lehre von den Punktmannigfaltigkeiten*, part 1 [Jahresbericht der Deutschen Mathematiker-Vereinigung, vol. 8 (1900)], part 2 [Jahresbericht der Deutschen Mathematiker-Vereinigung, IId supplementary volume (1908)], Leipzig.

Schoenflies, A., and Hahn, H., *Entwickelung der Mengenlehre und ihrer Anwendungen*, first half: *Allgemeine Theorie der unendlichen Mengen und Theorie der Punktmengen*, by A. Schoenflies, Leipzig and Berlin, 1913.

Key to Symbols

Index